CW01506928

The Development of The Faculty of Engineering

in the

University of Bristol 1909–2009

The Development of
The Faculty of Engineering
in the
University of Bristol
1909–2009

A History to mark its Centenary

R. T. Severn CBE, FICE, FREng.

Emeritus Professor of Civil Engineering
and Senior Research Fellow
University of Bristol

All rights reserved

No part of this publication may be reproduced, stored in a retrieval system
or transmitted, in any form or by any means, electronic, mechanical,
photocopying, recording or otherwise without the prior permission
of the author

Text copyright © R. T. Severn 2009
All illustrations © R. T. Severn or University of Bristol

ISBN 978-0-9561001-1-5

Designed, typeset, printed and bound by
4word Ltd, Bristol

Contents

The following short chapters, compiled from material supplied by them, record significant events and achievements of the six different Departments in the Faculty outside their contributions to the Faculty as a whole.

The author does not claim completeness in these records; he would be grateful for notice of important errors and omissions. As in the rest of this history, the date given is that of the year in which the session started.

Sons of Martha

Rudyard Kipling

A sizeable essay could be written on Kipling's use of the biblical sisters (John II) Martha and Mary, in his praise of Engineers as the providers of the facilities needed for many aspects of civilised life; but I will leave that to others. It is sufficient to say that, with poetic licence, Kipling exaggerates the difference between the two sisters; both were disciples of Jesus, with Martha being the more practical and questioning of the two.

> The sons of Mary seldom bother, for they have inherited that
> good part;
> But the sons of Martha favour their Mother of the careful soul
> and the troubled heart.
> And because she lost her temper once, and because she was
> rude to the Lord her guest,
> Her sons must wait upon Mary's sons, world without end,
> reprieve or rest.
>
> It is their care in all the ages to take the buffer and cushion the
> shock.
> It is their care that the gear engages; it is their care that the
> switches lock.
> It is their care that the wheels run truly; it is their care to
> embark and entrain,
> Tally, transport, and deliver duly the sons of Mary by land and
> main.

They say to the mountains , 'be ye removed', they say to the
    lesser floods, 'be dry'.
Under their rods are the rocks reproved – they are not afraid
    of that which is high.
Then do the hilltops shake to the summit – then is the bed of
    the river laid bare,
That the sons of Mary may overcome it, pleasantly sleeping
    and unaware.

They finger death at their gloves' end where they piece and
    repiece the living wires.
He rears against the gate they tend: they feed him hungry
    behind their fires.
Early at dawn, ere men see clear, they stumble into his terrible
    stall,
And hale him forth like a haltered steer, and goad him till
    evenfall.

To these from birth is belief forbidden; from these till death is
    relief afar.
They are concerned with matters hidden – under the earthline
    their altars are –
The secret fountains to follow up, waters withdrawn to restore
    to the mouth,
And gather the floods as in a cup, and pour them again at a
    city's drouth.

They do not preach that their God will rouse them a little
    before the nuts work loose.
They do not teach that his pity allows them to drop the job
    when they dam'well choose.
As in the thronged and lighted ways, so in the dark and the
    desert they stand,
Wary and watchful all their days that their brethren's days may
    be long in the land.

Raise ye the stones or cleave the wood to make a path more
    fair or flat –
Lo, it is black already with blood some Son of Martha spilled
    for that!
Not as a ladder from earth to Heaven, not as a witness to any
    creed,
But simple service simply given to his own kind in their
    common need.

And the Sons of Mary smile and are blessed – they know the
    Angels are on their side.
They know in them is the grace confessed, and for them are
    the mercies multiplied.
They sit at the Feet – they hear the Word – they see how truly
    the promise runs.
They have cast their burden upon the Lord, and – he has cast
    it upon Martha's Sons.

Acronyms and Abbreviations

AIC	Academic Innovations Centre.
ARP	Air-Raid Precautions.
AUT	Association of University Teachers.
BAC	Bristol Aeroplane Co. or British Aircraft Corporation.
BEng	Bachelor of Engineering.
BLADE	Bristol Laboratory for Advanced Dynamics Engineering.
BLWT	Boundary Layer Wind Tunnel.
BSc(Eng)	Bachelor of Science in Engineering.
CASE	Cooperative Awards in Science and Engineering.
CAT	College of Advanced Technology.
CATS	Credit Accumulation and Transfer Scheme.
CBI	Confederation of British Industries.
CEI	Council of Engineering Institutions.
CEng	Chartered Engineer.
CFD	Computational Fluid Dynamics.
CNAA	Council for National Academic Awards.
College	University College Bristol.
CVCP	Committee of Vice-Chancellors and Principals.
DEng	Doctor of Engineering – a postgraduate degree combining research and taught course elements.
DSA	Dept. of Science and Art.
DSIR	Dept. of Scientific and Industrial Research.
EC	European Commission.
ECTS	European Credit Transfer Scheme.
EEC	European Economic Community.

EERC	Earthquake Engineering Research Centre.
EMP	Engineering Management Partnership.
EngC	Engineering Council.
EngCUK	Engineering Council UK.
EPSRC	Engineering and Physical Sciences Research Council.
ERASMUS	European Community Action Scheme for the Mobility of University Students.
EU	European Union.
EuroMEng	MEng degree with study in Continental Europe.
Faculty	Faculty of Engineering, University of Bristol.
FCC	Faculty Computing Committee.
FREng	Fellow of the Royal Academy of Engineering.
FRS	Fellow of the Royal Society.
GCSE	General Certificate of Secondary Education.
HEFCE	Higher Education Funding Council England.
HESA	Higher Education Statistics Agency.
HoD	Head of Department.
IBM	International Business Machines.
ICE	Institution of Civil Engineers.
IChemE	Institution of Chemical Engineers.
ICL	International Computers Ltd. (a UK company).
IEE	Institution of Electrical Engineers.
IEng	Incorporated Engineer.
IGDS	Integrated Graduate Development Scheme.
IKB	Isambard Kingdom Brunel.
IMechE	Institution of Mechanical Engineers.
IP	Intellectual Property.
ISE	Institution of Structural Engineers.
JBEM	Joint Board for Engineering Management.
JIF	Joint Infrastructure Fund.
JREI	Joint Research Equipment Initiative.
LEA	Local Education Authority.
MASN	Maximum Aggregate Student Numbers.
MBE	Member of the Most Excellent Order of the British Empire.
MEng	Master of Engineering – an undergraduate 4-year degree.
MSc(Eng)	Master of Science in Engineering – a postgraduate degree.

MVB	Merchant Venturers Building.
MVTC	Merchant Venturers Technical College.
MVTS	Merchant Venturers Technical School.
NAAS	New Academic Appointments Scheme.
NAC	National Advisory Council on Education for Industry and Commerce.
OFFA	Office For Fair Access.
PAC	Public Accounts Committee.
PCFC	Polytechnic and Colleges Funding Council.
PEI	Professional Engineering Institutions.
Prof.	Professor.
PVC	Pro-Vice-Chancellor.
QAA	Quality Assurance Agency.
RAE	Research Assessment Exercise, Royal Academy of Engineering or Royal Aircraft Establishment, depending on context.
RAero Soc	Royal Aeronautical Society.
RAF	Royal Air Force.
RAM	Resource Allocation Mechanism.
RCUK	Research Councils UK (a joint organization of the 7 UK Res. Councils).
RE	Research Engineers – the students taking the DEng degree.
RED	Research and Enterprise Development (a University Organisation).
RIBA	Royal Institute of British Architects.
RWA	Royal West of England School of Architecture.
SAAS	School of Applied Social Studies.
SARTOR	Standards and Routes to Registration.
SAUS	School for Advanced Urban Studies.
SCUE	Standing Conference on University Education.
SERC	Science and Engineering Research Council.
SET	Science Engineering and Technology.
SME	Small and Medium Enterprises.
Society	Society of Merchant Venturers of Bristol.
SRIF	Science Research Infrastructure Fund.
SSRC	Safety Systems Research Centre.
SWRDA	South West Regional Development Assembly.
TCS	Teaching Company Scheme.

TEMPUS	Trans-European Mobility Programme for University Students.
TQA	Teaching Quality Assurance.
UCAS	Universities Central Admissions Service.
UCB	University College Bristol.
UCCA	Universities Central Council for Admissions.
UFC	Universities Funding Council.
UGC	University Grants Committee.
UIO	University International Office.
UMIST	University of Manchester Institute of Technology.
University	The University of Bristol.
UPARC	University Planning and Resources Committee
VC	Vice-Chancellor.
VP	Visiting Professor.
WMB	Wills Memorial Building.
WP	Widening Participation.

Important Dates

N.B.

1. The date given here, and throughout this history, is that of the year in which the session began.

2. Titles are dropped after first mention.

3. Professorial appointments are listed only when the holder subsequently became Dean; others may be found in Chapters 11–16. Assistant Deans are listed only to 1963; after this date the duties were divided between a number of academic staff; for example, by the late 1980s there were five Assistant Deans.

1909   MVTC and UCB settle their differences to create the Faculty of Engineering within the new University of Bristol.
Henry Overton Wills appointed first Chancellor.
Prof. Conwy Lloyd Morgan is the first VC, but is replaced by Sir Isambard Owen after 3 months.
Prof. Julius Wertheimer continues as Principal of the MVTC but is also Permanent Dean of the Faculty.
The University Charter day is celebrated on 24 May.

1912   Viscount Haldane appointed second Chancellor.

1919   UGC created as a Committee of the Treasury.
Prof. Andrew Robertson replaces Prof. J. Munro as Head of Mech. Eng.

1922   Dr. Thomas Loveday appointed third VC.

1924   Robertson becomes Permanent Dean and Principal of MVTC on the sudden death of Wertheimer.

1929   Winston Churchill installed as third Chancellor.

1940   King's College, London, Faculty of Engineering is accommodated by the Faculty.

1940	The 4th year of the Honours course is discontinued for the remainder of the war.
	Academic staff work at Bristol Aeroplane Co. during wartime vacations.
1943	Prof. David Robertson retires and is replaced by Prof. Gordon H. Rawcliffe as Head of Elect. Eng.
1944	Prof. Alfred G. Pugsley appointed Head of Civil Eng.
1945	Prof. Arthur M. Tyndall (Physics) becomes acting VC.
1946	Sir Philip Morris appointed VC.
	Bristol Aeroplane Co. gives Univ. £6000 p.a. for 10 years for the creation of a Dept. of Aeronautical Eng. and the Sir George White Chair in that Dept.
	Prof. A. Roderick Collar appointed to the Chair.
	John L. M. Morrison appointed to a Chair in Mech. Eng.
1950	The Society of Merchant Venturers relinquishes its direct responsibility for the Faculty.
	Andrew Robertson retires, and Morrison becomes Head of Mech. Eng.
	Rawcliffe becomes Dean for a 3-year term. (N.B. from here forward, the Deanship was for a fixed term, which until the end of the century was understood to be 3 years).
	Dr Ernest F. Gibbs (Civil Eng.) appointed first Assistant Dean to deal with transfer from MVTC to New Building.
1951	School and Higher School Cert. replaced by General Certificate of Education (O- and A-levels).
1952	Rawcliffe ill; Pugsley becomes Dean.
1954	Collar and S. Toby Newing (Eng. Maths.) become Dean and Assistant Dean, respectively.
1956	Pugsley knighted.
1957	Morrison becomes Dean.
	British Computer Society founded.
1958	H. M. Queen Elizabeth II opens Queen's Building on 5 December.
	Munich aircrash – Collar appointed to Committee of Enquiry.
1959	Frederick de La C. Chard replaces Newing as Asst. Dean.
	W. M. Shepherd (Theor. Mech.) appointed to a Chair.
	First discussions on Royal Western Academy (RWA) School of Architecture joining the University.
1960	Shepherd becomes Dean (1960–63).

1961   Pugsley appointed PVC (1961–64).
       The Science Research Council (SRC) is created.

1962   Pugsley becomes Acting Vice-Chancellor due to the illness of Philip Morris. Gordon F. C. Rogers (Mech.) becomes Asst. Dean.

1963   RWA School of Arch. integrated with Univ. to form Dept of Arch.
       Rawcliffe elected Dean.
       Great Hall in Wills Mem. Building reopened after repair of war damage.
       Publication of the Robbins Report on Higher Education.

1964   Prof. John Harris (Botany) appointed Vice-Chancellor from August 1965.

1965   Faculty's first MSc course, in 'Fluid and Thermal Studies for Industry'.

1966   Duke of Beaufort installed as fourth Chancellor.
       Collar appointed PVC.
       Rogers becomes Dean.
       Intermediate course discontinued.

1967   Jones ('Brain Drain') Report.

1968   Dainton Report on Science and Technology in Higher Education.
       Collar becomes VC for 17 months on the death of John Harris.
       Pugsley retires and is replaced by Roy T. Severn as Head of Civil Eng.
       Pugsley appointed to Ronan Point (London flats collapse) Enquiry.

1969   Prof. Alec W. Merrison (Univ. of Liverpool) becomes Vice-Chancellor.
       Shepherd appointed Dean.

1970   Appointments Board becomes Careers Advisory Service (CAS)
       Merrison is appointed Chairman of the Steel Box-Girder Bridge Enquiry.

1971   Prof. Dorothy Hodgkin O.M., FRS installed as fifth Chancellor.
       Shepherd retires and Severn becomes Dean.
       Morrison retires and is elected President of IMechE.
       Kenneth F. Sander (Trin. Coll. Camb.) appointed Prof. of Electronic Eng.
       Ronald D. Milne appointed Prof. of Theor. Mech.

1972   UK joins the European Community.
       Colin Andrew appointed Prof. and Head of Mech. Eng.
       Rawcliffe elected to Royal Society.
1973   Dept. of Theor. Mech. becomes Dept. of Engineering
       Mathematics.
       Dept. of Electrical Engineering becomes Dept. of Electrical and
       Electronic Engineering.
       Collar retires and is replaced by Lewis F. Crabtree from RAE
       Farnborough.
1974   Sander appointed Dean.
1975   The three senior Engineering Institutions (Civil, Mechanical,
       and Electrical) set up Committees to review Routes to
       Membership.
       Rawcliffe retires.
1976   Fellowship of Engineering founded. Pugsley, Collar, Morrison
       and Rawcliffe are founder members.
       Merrison becomes Knight Batchelor.
1977   Andrew Robertson dies aged 94.
       Finniston Committee established; it publishes its Report in
       1979.
       Degree course in Eng. Maths. started.
       Rogers appointed PVC and Andrew becomes Dean.
1979   B. Michael Bird appointed Prof. of Electrical Eng. and Head
       of Dept.
1980   Rawcliffe dies.
       Milne becomes Dean.
1981   Science Research Council becomes Science and Engineering
       Research Council.
       Severn appointed PVC.
1983   The Fellowship of Engineering receives its Royal Charter.
       Crabtree becomes Dean.
1984   Department of Architecture closed.
       Sir Alec Merrison retires, and Prof. P. Haggett (Geog.) becomes
       acting VC.
1985   Sir John Kingman (Chairman of the SERC) becomes VC.
       Geology moves to Wills Memorial Building.
       First Faculty Research Centre created in Earthquake
       Engineering.
       First Faculty MEng degree starts in Mech. Eng.

Jarratt Ctte on Efficiency Studies in Universities reports.

Government Green Paper on Higher Education into the 1990s.

University Resource Allocation Mechanism (RAM) introduced.

Bird becomes Dean.

Eric L. Dagless appointed to the new Imperial Group Chair in Microelectronics.

Prof. Joe H. McGeehan appointed to the Chair in Communications Eng.

1986   Aeronautical Eng. changes its name to Aerospace Eng., but its degrees are still in Aeronautical Eng.

First University Grants Committee Research Assessment Exercise (RAE).

University Research Committee created.

1988   Universities Funding Council (UFC) replaces UGC for the funding of 52 universities.

Polytechnic and Colleges Funding Council (PCFC) created to fund 50 Polys. and Colleges.

Dagless appointed Dean.

1989   Computer Science Dept. transfers to Engineering from Science Faculty.

Sir Alec Merrison dies.

University starts 'Modularisation' discussions.

The second Research Assessment Exercise is held.

David I. Blockley appointed to a Personal Chair in Civil Eng.

1990   Sir Jeremy Morse becomes sixth Chancellor.

Undergraduate Prospectuses now produced by University rather than by individual Faculties.

Dr. Alisdair Lockhart appointed University Development Director.

The 4-year MEng course with Study in Continental Europe started.

Bird appointed PVC.

1991   Campaign for Resource officially launched.

Severn becomes Dean.

The Food Refrigeration and Process Eng. Res. Centre (FRPERC) created.

A new building for the Faculty is announced.

1992   The Fellowship of Engineering adopts the title of The Royal
        Academy of Engineering.
        HEFCE created by merger of UFC and PCFC to remove the
        'Binary Divide'.
        The third Research Assessment Exercise is held.
1993   HEFCE introduces Teaching Quality Assessment (TQA).
1994   Blockley becomes Dean and introduces a Faculty Resource
        Allocation Mechanism (FRAM).
        Engineering and Physical Sciences Research Council replaces
        SERC.
        University joins Russell Group of Research Universities.
        Construction starts on a new building for the Faculty.
        University appoints a Deputy VC – Prof. Brian Pickering
        (Anatomy).
1995   First issue of 'The Brief'– a Faculty newsletter.
        Safety-critical Systems Research Centre (SSRC) started.
        Severn retires and is replaced by David Muir-Wood from
        Glasgow Univ.
1996   HRH the Duke of Edinburgh opens the Merchant Venturers
        Building (MVB). Fourth Research Assessment Exercise.
1997   McGeehan becomes Dean for a 6-year term.
        Sir Ron Dearing's Report on 'Higher Education in the Learning
        Society' is published.
        Quality Assurance Agency (QAA) created with funding from
        HEFCE and Universities.
1998   Telecommunications Research Centre established by Toshiba
        Corpn.
1999   Student Engineering Society re-born after 2-year in
        abeyance.
        RED – the University Research and Enterprise Development
        Centre founded.
2000   The proposed Bristol Laboratory for Advanced Dynamics
        Engineering (BLADE) receives £15 million from HEFCE and
        the Office for Sci. and Tech.
2001   Kingman retires and Prof. Eric Thomas (Southampton)
        becomes VC.
        MEng in Engineering Design (RAE-sponsored) started.
        Fifth Research Assessment Exercise.
        An EPSRC 5-year postgraduate EngD degree is launched in

Non-Destructive Evaluation, in partnership with 6 other UK universities.

2002 BLADE reconstruction of Queen's Building starts.

University 'Campaign for Resource' ends.

Nick Lieven (Aerospace Eng.) awarded a Personal Chair.

2003 The Chancellor, Sir Jeremy Morse, lays the BLADE foundation stone.

Muir-Wood becomes Dean for a 4-year term.

Roberts review of the Research Assessment Exercise.

Prof. David Smith (Mech.) appointed Faculty Director of Research.

2004 The Rt. Hon. the Baroness Brenda Hale of Richmond becomes the seventh Chancellor.

Faculty laboratories reorganised into 3 Faculty groups with an overall Technical Services Manager.

Govt's Higher Education Act introduces 'Widening Participation'.

2005 H. M. Queen opens BLADE. Faculty organises research into 6 cross-departmental themes.

2006 Careers Advisory Service becomes Careers Service.

Two EPSRC 5-year postgraduate EngD degrees started; one with Univ. of Bath in Systems Engineering, and one with 6 other UK universities in Nuclear Eng.

2007 Lieven becomes Dean for a 5-year term.

Prof. John Hogan (Eng. Maths) awarded £4m to establish a Faculty Centre for Complexity Science.

The Advanced Composites Centre for Innovation and Science opened by the Minister for Innovation and Science.

2008 The sixth Research Assessment Exercise.

Prof. David Clarke (Law) appointed Deputy Vice-Chancellor.

Illustrations and Tables

Illustrations

Tables

Acknowledgements

In compiling this history of the Faculty of Engineering I have received much help and assistance from many individuals and Departments of the University. If I were to list all these sources of information it would take up much space, and, even so, there would be an unforgivable possibility of forgetting some; I have therefore refrained from doing so. Instead, I have opted to thank all those members of the various organizations within the University which have given me their time and knowledge, and I sincerely hope that the individuals concerned will recognize themselves and will accept my gratitude.

Throughout the two years spent writing this book, I have received unstinting help from the various parts of the University Library, particularly the Special Collections section, who spared no efforts in finding for me fascinating documents and photographs of the early years (it is with thanks that I record the use of photographs in DM508 (Queen's Building envelope), DM1978/1/38 and DM615). To the staff of the Queen's Building Library I owe especial thanks for their forbearance and unfailing kindness in providing all manner of Minutes of Faculty meetings, Prospectuses and Dean's Reports from 1909 to the present day, and also providing me with a quiet space in which to write parts of this book.

As I wrote in the Preface, many aspects of the Faculty's history have been linked to that of the University, and here the help given to me by all sections of the Administration in Senate House is noted, in such areas as student fees and statistics, exchange schemes with European universities, and data concerning the Research Assessment Exercises.

The Registrar's own library has been made available to me for Senate and Council papers.

Within the Faculty itself, my colleagues have always been willing to either support, or correct, my own memory of events in which we were both involved, and for Chapters 11–16 I give my thanks to Heads of Departments and their staff for collecting the information recorded there.

I have been conscious in writing this History of an inability to give proper acknowledgement to the contributions made to the progress and success of the Faculty by the non-academic staff. In all University faculties it is the administrative staff who hold the fabric together, but in the Engineering Faculty I have always been aware that it is the technical staff who held an essential key to success in both teaching and research. Regrettably, the records do not often list either their names or their contributions.

The only individual whose name I will mention is that of my wife, Hilary. She has greatly enjoyed with me, more than fifty years of belonging to the University, never once complaining that it was the only rival she ever had in my affections. She has happily played a full part in my academic life, and her ability now to recall dates, names and places, has been of great help in writing this History.

Preface

Although this book aims at being an account of the hundred years of development of the Faculty from its hesitant beginning in 1909 to the present day, its conception and gestation requires me to go back much further. To be specific, its development can only be imperfectly understood without reference to the Society of Merchant Venturers of Bristol, the Technical College which it created in 1894 and its earlier Technical School of 1885. But even that school had its origins in two distinct streams of technical education in Bristol, one being supported by the Society since at least as early as 1737, and the other by the Bristol diocese since 1821. There were others, such as the Government School of Science and Art (1852–1936), but since they seem to have had no influence on the development of the Faculty, they do not appear again in this history.

I have therefore used the first three chapters to describe these early activities before coming to the century of development of the Faculty within the University of Bristol. In doing so I have had to face a matter of terminology, because I believe it to be necessary to understand the distinction between 'technical education' on the one hand and 'engineering education' on the other. In my own personal dictionary the former term is defined as the transmission of known specialist information relating to a particular human activity – each one of which has its own technology – whereas the latter, though containing some elements of the former, is essentially concerned with principles, the application of which can lead to new activities in the broad subject of engineering. In other words, it is the university concept of 'teaching within an atmosphere of research'. The distinction in terminology which I make here is not entirely a personal one within the context of this history, because in a study of documents leading to the creation of University College Bristol in 1876, one finds its first Principal, Alfred Marshall, a political economist by profession, saying '. . . It is expedient to develop the teaching which the College gives in the *scientific principles of engineering* (the italics are mine).

He may have had in his mind, as I do, that engineering education of this kind, which in proper hands leads to both numeracy and literacy in the individual, has value as a precursor to success in many professions. The employment routes of a sizeable proportion of our recent Bristol engineering graduates, and their successes in such professions as banking, insurance and commerce gives testimony to this fact.

The lack of clarity in terminology has not entirely disappeared because some respected organisations, and even government departments, whilst distinguishing between science and technology, imply that the latter is synonymous with engineering. Fortunately, the Royal Society and the Royal Academy of Engineering are not guilty in this regard, both using what I consider to be the more appropriate, and precise, phrasing of 'Science, Engineering and Technology'.

The reader of this history will discover that I have found it necessary to spend time in referring to matters which affected not only the Faculty but the whole University. The justification for this is that they all had an appreciable affect on the development of the Faculty, which could not be explained without reference to the wider University picture. As a most significant example, the policy of almost all governments since the 1950s, both Conservative and Labour, has been to encourage the growth of university studies in all disciplines, and particularly in science and engineering; but it was an encouragement only recently backed up by the provision of the necessary resources; the universal political cry was 'rationalisation and efficiency', or 'more for less'. As we shall find, in the early 1980s these financial pressure caused the University to close the School of Architecture; not only had the Faculty played a major role in creating and establishing it within the University, but it had a significant part to play in events leading to its closure. In the same period the Faculty itself accepted acutely diminishing resources – which inhibited its progress – as a contribution towards ensuring that the University as a whole was not seriously damaged by the cuts in resource imposed by the government. In truth, it was not until the New Labour government in the early years of this century began to provide the necessary financial support, that the University and Faculty were able to begin to realise their full potential.

Another matter concerning the structure of this book is what I believe to be common practice in historical texts. I have dedicated each chapter to a reasonably well-defined period of years, and within this period, topics begin, but do not always end. I have therefore usually adopted the practice of continuing with a full description in the chapter in which the topic begins, rather than disrupting the continuity by holding developments to one, or more, succeeding chapters.

On a personal note, after being a member of the Faculty for fifty-two years, and having served it in various capacities from lecturer to pro-vice-chancellor, I must comment that I have derived great pleasure from writing this history, and in re-living some of the events in which I played a part. I had the privilege of knowing Andrew Robertson, who really gave the Faculty its initial steer towards excellence in both teaching and research, and of the Gordon Rawcliffe, Alfred Pugsley, Roderick Collar, John Morrison and William Shepherd quintet of professors who, after the Second World War, created a Faculty of international excellence. Their successors have done much to extend the breadth of the Faculty's activities whilst continuing to maintain its standing within the top group of UK universities, and in some areas establishing *primus inter pares* internationally. Under pressure from outside bodies, such as the Engineering Council, the Royal Academy of Engineering and the Professional Institutions, they have sometimes been forced in their undergraduate courses to move away somewhat from Alfred Marshall's 'scientific principles of engineering' – a phrase which was repeated until 1967 in the Faculty Prospectus describing the basis of these courses. Under the umbrella title of 'The Engineer in Society', such topics as law, politics, finance, management and even the newly discovered topic of 'Entrepreneurship' have found their way into the undergraduate curriculum, which, for these and other reasons, now occupies four years for the majority of students. In fact, it could be said that universities are now being required to play a greater part than before in the 'formation' of the professional engineer, with industry correspondingly less – the general area of 'design' being a particular example. Only the future of the UK engineering profession will tell whether such changes have been beneficial.

One theme running through this history of the Faculty is the increasing part which governments have played in controlling university education, both in terms of number of students and in directing which subjects they should study. Necessarily, this involved financial resources, the amount of which was sometimes determined, for good or ill, by political beliefs. At no time in the hundred years were universities totally independent of what ultimately was government support, whether it came direct, from research councils or from local authorities, and as the years passed the dependence increased to what many consider to be an academically unhealthy level; it sometimes brought the satisfaction of political aims, rather than the increased realisation of international quality in teaching and research. At the end of the century covered by this book, it is comforting to see amongst the more prestigious universities, such as the 'Russell Group', attempts to restore a greater measure of financial freedom through the American model of self-help and support from former students and charitable bodies.

If I have a concern for the future of the Faculty, it is that its growth has meant fewer contacts between its departments, and in a wider sense, in the growth of the University we have lost the educational and human contacts with departments and colleagues in other faculties, which I found to play such a significant part in my general education and enjoyment of my half-century at Bristol. This aspect of the corporate life of the University – initiated in 1928 by the action of two professors who proposed the establishment of a luncheon club for academic staff – developed during the next 50 years into a self-financing Senior Common Room, which apart from providing meals of various kinds in its own dining rooms, organised a wide variety of activities which brought together colleagues from different departments. Sadly, growth has meant that such activities now occur, if at all, at faculty or departmental level, with the result that in 2008 the Senior Common Room was deemed to have no purpose and was closed down.

Chapter 1

The Society of Merchant Venturers and the Bristol Diocesan Schools, 1737–1885

1.1 Introduction

During this 150-year period there were, at different times, three organisations in Bristol concerning themselves with training and education for engineering. The two schools which were involved in training for local industry, and which eventually merged in 1885, were that supported by the Society of Merchant Venturers and that supported by the Church of England in its Bristol Diocese. The third organisation was that provided by the creation of University College Bristol (UCB), which provided education and some training in several branches of engineering. But since UCB was not created until 1876, which is towards the end of the period being discussed here, it will be convenient to deal first with events leading up to the merger of the Society and Diocesan schools in 1885 to form the Merchant Venturers School, before considering developments at UCB in the following chapter.

There is some evidence (Ref. 1) that the Society – founded in 1552 – had a vocational school as early as 1595 for providing a supply of men to sail and manage their ships, but for our purposes the creation of its Mathematical and Navigational School in 1737 can be taken as the starting point. It used legacies left to them for the purpose of teaching young men the skills needed by the mariner, and it is probable that it was situated within the Society's own hall, then in King's Street. In 1783 the Navigation School was combined with what was referred to as the Writing School, a general school of longer history, also run by the Society. But when the merged school was inspected in 1792 it was found that no boy had been instructed in either mathematics or navigation, making it clear to the inspector that the available funds had not been used for their intended purpose. As a partial defence of this failure, it is to be remembered that the role of Bristol as a major port had already begun to diminish, resulting in less demand for maritime employment. Despite the inspector's adverse report the school continued to function, but by 1839 it was again considered to be inefficient, leading to amalgamation with a school which taught navigation under the auspices of the Bristol Marine Society. But history repeated itself to the extent that very few boys were studying navigation, resulting in the Society disengaging itself from the Bristol Marine Society in 1844.

1.2 The Influence of Canon Moseley and Lyon Playfair

It is now convenient to turn to the second stream of contribution to training for industry in Bristol. It came, in due course, from the founding in 1821 of a Diocesan school for educating poor children in the principles of the Established Church, but by 1854 it was considered to have lost its purpose due to the system of national and parish schools which had been created throughout England and Wales. When the Diocesan Committee met to consider its future, it sought the advice of Henry Moseley, then Canon of Bristol Cathedral, but of greater importance, he had previously (1831-44) been Professor of Natural and Experimental Philosophy and Astronomy at King's College London, an academic establishment where it was usual at that time to combine ordination in the Established Church with other academic interests. To understand the significance of, and reason for, this invitation to Moseley, two factors must be recorded.

It is normally assumed that it was Prince Albert's[1] sponsorship of the idea of holding the Great Exhibition of 1851 in Hyde Park, having the objective of informing the world about the excellence of British achievements during the early parts of what is usually referred to as the Industrial Revolution. But although it was housed in William Paxton's truly innovative and prefabricated structural creation of cast iron and glass[2] – the 'Crystal Palace' as it came to be called – the really

---

[1]   Prince Consort to Queen Victoria, 1819–61.

[2]   Columns and beams were of cast iron; wrought iron was used for wind bracing and ties. Timber was also used extensively. When it was moved from Hyde Park to a site in south London, the location took the name 'Crystal Palace'.

significant effect of the Great Exhibition was to alert the nation in general, and to Moseley in particular, to the fact that many countries, notably those in Europe, had made advances which were superior to our own in many branches of industry and in types of manufacture. Whereas our manufactures were still craft-based, others had applied the results of scientific research to improve both the quality and quantity of the goods which they produced.

The second factor leading to Canon Moseley's invitation to advise the Diocesan Committee was that Bristol itself had created many new industries, notably paper, sugar and tobacco, to replace its diminishing maritime importance, but had no existing facilities for the formal training of its citizens for employment in these industries. Moseley would have known that such facilities were being considered in Europe and in cities in the north of England, and accordingly suggested to the Diocesan School Committee that its Day School should be re-styled as the Bristol Diocesan Trade and Mining School (1856–75) which would be

> ... similar to those in Germany and other countries wherein a
> special course of instruction of a practical kind, having
> reference to the mechanical and practical callings and the
> trade of Bristol might be introduced.

His aims were clear, because he went on to say '. . . that it would be necessary to secure the services of persons skilled in the practical application of the sciences to be taught'. Also behind Moseley's advice would have been the knowledge that the Great Exhibition had resulted in the creation of a government Department of Science and Art (DSA, 1853) with responsibilities for secondary and technical education and financial resources to distribute to organisations taking part in these activities, whilst the existing Department of Education concerned itself with primary education. It is to be noted here that the word 'Art' as used in the DSA title meant more than the graphic arts – the creation of paintings and sculptures; it meant, to use one of the interpretations given by the Oxford Dictionary, the 'practical application of any science, industrial pursuit or craft'. Such a definition had been used one hundred years earlier when the Royal Society of Arts, Manufactures and Commerce received its Royal Charter. The Diocesan Committee accepted Moseley's suggestion, and at its opening in 1856, Dr. Lyon Playfair, one of the Joint Secretaries of the DSA '. . . congratulated the

people of Bristol on being the first in England to contemplate the establishment of such a valuable institution'. Historians of science will recall
that Dr Lyon Playfair, essentially a chemist, played a vital role in the
early promotion of English science, and science education. As a fellow
German speaker (his PhD degree was from a German university), he
was invited to assist the Prince Consort in his planning of the Great
Exhibition and this led to the Prince having influence on Playfair's
subsequent appointment as an Exhibition Commissioner[3]. Of interest
to our history is that in 1852 he published an essay *On Technical
Education* in which he was probably the first to interpret the decline
of British industrial prestige to a national deficiency in technical
instruction. In 1858 he became Professor of Chemistry at Edinburgh
University, followed by election to Parliament as Member for the
Scottish Universities, and in 1885 he was elected President of the British
Association for the Advancement of Science.

1.3 The Bristol Trade and Mining School

As well as a preparatory department in the new Trade and Mining
School for those aged between 9 and 13, the secondary department concerned itself with distinct streams in either commercial studies or
mathematics and applied sciences. For adults there were day classes in
chemistry, mining (for the important Bristol and Somerset coalfields)
and engineering, although precisely what was taught in the last of these
subjects is not known. There were also evening classes in a wide range
of practical topics.

In 1863 the Society of Merchant Venturers, having observed the
success of the Trade School, made a financial arrangement with it for
the teaching of navigation, but this proved to be unsatisfactory causing
the Society to reopen its own school, but the demand for knowledge of
navigation was not forthcoming and it was finally closed in 1878.

Although the mining interest of the Trade School was lost in 1868,
its inspection by the Endowed Schools Commission in that year
resulted in an excellent report, but even so, by the early 1870s it was
experiencing financial difficulties. There is no doubt that members of

---

[3] The Royal Commission for the Great Exhibition of 1851 was incorporated by supplemental
charter as a Permanent Commission after winding up the affairs of the Great Exhibition. It
still exists, with HRH the Duke of Edinburgh as its President, using funds derived from its
South Kensington properties to make prestigious awards to young scientists and engineers.

the Society would have drawn great benefit in their various business activities from the skilled workforce produced by the Diocesan Trade and Mining School, and it is therefore not surprising that at this time of financial difficulty for the school, the Society were willing to take over the total costs and administrative control, and to continue to run it on the well-established and successful lines; the only change being the removal of the word 'Diocesan' from its title. It should be made clear, however, that the Society were undertaking to cover recurrent costs only, and did not at this stage attempt to own the Trade School build-ings; these belonged to the governors of the Colston[4] School Trust.

1.4 The Merchant Venturers Technical School and Technical College

The Bristol Trade School, as it now was, continued to flourish, so much so that in his 1879 Report, the inspector from the DSA drew attention to the fact that large class numbers made the available accommodation insufficient, requesting that action be taken to remedy the situation. Because the Society was now responsible for running the Trade School, and action consequently lay with them, it instructed its Clerk to look for suitable sites. It was certainly fortunate that shortage of space had also become a problem for Bristol Grammar School which was about to vacate its premises in Unity Street following a move to Tyndall's Park – a site which it continues to occupy – and in 1880 the Society pur-chased the Unity Street site, informing the Colston School trustees of its intention to build a new home for the Trade School, coupled with the possibility that it would offer it to the trustees at a nominal rent.

It was at this stage of completion of the new building that the great value which the Society placed on the Trade School became clearly apparent, because on being informed that the financial accounts of the Colston Trust showed that they could not support all their commit-ments, which included the rent of the Trade School's new building, the society offered

---

[4] Edward Colston (1636–1721) – a wealthy London merchant but born in Bristol, bequeathed his great fortune to good causes, making the Society trustees of some of them, including those concerned with education of both boys and girls.

> To take upon themselves the entire charge of the School . . .
> and that from the period of taking over the School, the whole
> expenses of carrying it on be borne by the Society . . . ; and
> that it is the intention of the Society to develop all the
> branches[5] of the School . . . for the promotion of scientific and
> technological teaching.

In this manner the Merchant Venturers Technical School (MVTS) was founded in 1885 at a total cost to the Society estimated to be £40,000, of which three-quarters was for the new building. As a consequence of this munificence the Colston School Trustees were able to fulfil a commitment which they had accepted in 1875 of creating the Colston Girls School.

The Society enthusiastically supported and developed its new school, initially through the incumbent headmaster, Thomas Coomber, who had been in that post since 1856, but it seems clear that the Society had ambitions for the School which he was not capable of realising. Means were therefore found for easing him into retirement through the inducement of a pension of £300 p.a., which at that time was more than generous! His successor in June 1890, at a salary of £500 p.a., but with no provision for a pension, was Julius Wertheimer, a man who was to play a major role in the early years of our Faculty, becoming its first Dean in 1909. He was of an altogether different character from Coomber as later events were to prove. He had obtained a BA degree at University College, Liverpool and a BSc in chemistry at Owens College, Manchester, and his previous appointment had been as head of Leeds College of Science and Technology. His first acts at the MVTS were indicative of the aggressive, but successful and productive style of governance for which he became well-known subsequently in his dealings with University College Bristol. He was critical of almost all aspects of the MVTS – its discipline, the intellectual level of the courses, and the capabilities of existing teachers – although he only succeeded in dismissing one of them. He was however the type of new broom which the MVTS needed and which the Society wanted, because from an enrolled student body for day and evening classes of 1384 in 1890, he raised it

---

[5] This meant primary, secondary and technical education. It was not until 1908 that the primary element was transferred to Bristol Grammar School.

to 2065 in 1896 and to 2512 in 1903. One aspect of his character was, however, detrimental to the eventual well-being of the MVTS; he was extremely jealous of his autonomy as Principal, not accepting the possibility of interference from any other person or organisation. This arose acutely in 1891 when the City of Bristol established a Technical Instructions Committee[6] with a budget of £5700 p.a, to be distributed to suitable bodies for the expansion of technical education, but carrying with it the requirement of representation on governing bodies. The MVTS declined to bid for such funding although they did accept some scholarships for their students. It appears from the records that Wertheimer and his MVTS benefited considerably from other sources of funding; from Central Government, and particularly from the Society itself, which from 1885 had been prepared to spend up to £2000 p.a. on what in 1884 had become a Technical College (MVTC) rather than a School. The justification for this change of title, indicating the ambitions of both Wertheimer and the Society, was that 'College' indicated a more important place in the educational spectrum and 'Technical' indicated the type of education provided. By this change it is more than probable that Wertheimer was beginning the confrontation with University College Bristol, founded in 1876, which is discussed in detail in the next chapter.

The change of name from MVTS to MVTC was not only accompanied by an increase in student numbers during 1890–1903, but also by an enhancement of the level of instruction given, from a basic technician level to BSc courses for external examinations of the City and Guilds of London, and the University of London, in almost all existing branches of engineering. These changes were responding to the need for technical education in the Bristol area, a need which some other institutions took part in filling. There was, for example, the short-lived Bristol School of Practical Art, founded in Queen's Road in 1853 under the auspices of the Board of Trade, but of far greater importance was the development of engineering studies in University College Bristol, and it is to this that we must shortly turn. But before doing so, it will be instructive to trace corresponding developments in Europe, as well as the effects of the Great Exhibition of 1851.

---

[6] The Technical Instruction Act of 1889 enabled local authorities to raise a penny rate in support of technical instruction.

1.5 Early Developments in Europe

In view of the priority which must be given to France, later followed by Germany and then Russia, in the development of engineering education, it is interesting that, as in Bristol and the rest of the UK, in France it progressed from *training* to *education* as a matter of necessity. As early as 1720 the French government, following its less than successful War of the Spanish Succession (1702–13), established the Corps of Engineers of Ways of Communications, and in 1747 the *École des Ponts et Chaussées* was founded in Paris for training military engineers in the construction of roads, canals and bridges. But the French Revolution of 1779 and war against the European Powers produced the need for new ideas in the construction (and destruction) of the same roads, canals and bridges, resulting in the creation of what is now the most prestigious *Ecole Polytechnique* (Ref. 2). Its organisation was very different from existing schools, in that entrance was made available to all social classes and was subject to success in a very competitive examination. Its founder, the esteemed mathematician and engineer Gaspard Monge, was soon joined by men such as Lagrange, Fourier and Poisson, names which continue to resonate in the minds of engineering educators of today. In this new *École* the teaching system was quite different from that of the older schools, which had been based, as in the UK more than 100 years later, on the apprenticeship principle, in which established engineers explained to one or more students how different types of structure were actually designed and built. In the *Ecole Polytechnique*, the now familiar lecture system to a group of students was established, with the contents consisting of general scientific subjects such as mathematics, mechanics, physics and chemistry. It was accepted as a principle by these men, that if the student received a good education in the basic sciences, it would not be difficult for them to acquire the specialist knowledge required for the different branches of engineering. As time progressed, the *Ecole Polytechnique* drifted more towards the fundamental sciences and became, amongst other activities, the provider of the two-year first stage of engineering education for the very brightest of students in France, to be followed by three years of more professionally oriented education (and some training) at one of the relevant *Grande Ecoles.*

In the period after the Napoleonic wars the development of engineering education in the German states[7] also began, the years 1815–40 seeing Polytechnic Institutes appear in many cities. But their impact was significantly different from that in France, in that the direct interaction with industry was very strong, no doubt contributing to the speed with which the German states surpassed other European countries in the quality and volume of its manufactured projects during the years immediately before Prince Albert's Great Exhibition of 1851.

As far as the origins of Russian engineering education are concerned, all that need be said is that its first appearance was in the St. Petersburg Institute of Engineers of Ways of Communications, founded in 1809 by the French engineer Betancourt, who was also its first director. The teaching methods were therefore those of the *Ecole Polytechnique*, but since its teachers included Lame and Clapeyron, who actually designed and built cathedrals and suspension bridges in St Petersburg, there were likely to have been elements of training also in its contents.

1.6 The Great Exhibition of 1851

My own interpretation of developments is that *technical* education for industry in England, as I define it and as practised in the Merchant Venturers Technical School, was found wanting by the Great Exhibition of 1851, and that those responsible for the creation of University College Bristol in 1876 were intent on following the lead given by France and Germany in providing *engineering* education based not only on mathematics and applied science within a research environment, but also within a larger environment of the multi-disciplinary organisation which a university provides. When the Merchant Venturers school became a college in 1894, it had already absorbed some of the lessons from the Great Exhibition, but not sufficiently to avoid the

---

[7] The German states were not fully unified until 1871 by the political skills of Bismarck and the military skills of Moltke. For the latter, *The Times* obituary records that his success was based upon the employment of three recent engineering developments – the breech-loading rifle, the electric telegraph and the new railway system in Europe. The telegraph was used to keep in touch with his field officers, and the railways enabled him to calculate the speed with which he, and his enemies, could change the disposition of their forces.

significant controversy leading up to the creation of the University of Bristol in 1909 in which the Society of Merchant Venturers eventually agreed to merge the higher level activity of its college with the Department of Engineering in University College Bristol, thereby creating the Faculty whose centenary this book celebrates.

Chapter 2

Engineering at University College Bristol, 1876–1909

2.1 Early Proposals for the College

Details of the founding of UCB in 1876 are dealt with fully by J. G. Macqueen and S. W. Taylor (Ref. 3), but it is noted here that the original intention of its sponsors was to create a Bristol College of Science, and that in 1873 a committee chaired by Gilbert Elliot, Dean of Bristol, proposed that a 'Technical College of Science' be established. But a year later, largely through the efforts of the Revd Dr John Percival, Headmaster of Clifton College and Dr Benjamin Jowett, Master of Balliol College Oxford, it was agreed at a public meeting that the new foundation should be a 'College of Science and Literature for the West of England and South Wales, to be established in cooperation with certain Colleges in the University of Oxford,' with the already established Bristol Medical School (1833) as a constituent part. Professors of engineering and of mining were to be appointed, the former responsible for descriptive geometry, mechanical philosophy and mechanics, whilst the latter's remit was in geology, mineralogy, the art of mining and surveying. Jowett's view of what the College should be, is expressed in the following quotation, taken from the speech he made in the Victoria Rooms, Bristol, on 11 June 1874,

> so there are other studies which will hardly succeed except in the great centres of industry. The professions of medicine and engineering have their natural homes in large towns ... No man will be a first rate physician or engineer who is not

something more than either, who has not some taste for art, some feeling for literature, or some interest external to his profession.

It is useful to recall here that in the other stream of education in Bristol which was described earlier, the Diocesan Trade and Mining School had been founded in 1852 and substantially augmented in its activities by the Society of Merchant Venturers in 1875. It is not recorded what Gilbert Elliot himself meant by 'Technical College of Science', but clearly he was dissatisfied with what already existed in Bristol. There could have been personal, religious or political reasons, which, as we shall see, certainly presented themselves during the next thirty years, but it is also possible that his intention was to follow up the evidence of the 1851 Great Exhibition by creating a college in which the educational purpose was in the application of science for the practical purpose of trade and manufacture, which he construed to be engineering. If this is so, there are few professional engineers today who would accept Elliot's view that engineering is simply an applied science. Rather, they would applaud the Jowett/Percival direction that UCB should dedicate itself to both arts and sciences, so that, in theory at least, education which concentrated in either of these two areas would be carried out in an environment which included teachers and students in the other.

The first day of the first session of UCB was 10 October 1876, and early Calendars give an introduction to its purpose and function in the following terms:

> The College supplies for persons of either sex above the ordinary School age the means of continuing their studies in Science, Languages, History, Literature and Theory of Music and particularly affords appropriate and systematic instruction in those branches of Applied Science which are more nearly connected with the Arts and Manufactures. Courses of instruction have been specially arranged for students intending to become Engineers, Surveyors or Architects. Special attention is given to class teaching and laboratory work.

The above quotation has been included for three reasons, the first being that UCB was providing an education in applied science for those intending to become engineers, surveyors or architects, and not technicians for immediate employment. In view of what was to happen some 80 years later, the second reason for including the quotation is the inclusion of architectural education, and the third is that there was no mention of either postgraduate studies or of research, although these were included much later in the Calendar for 1906–07 and subsequently.

2.2 A Department of Engineering within the Faculty of Science

The creation of a Department of Civil and Mechanical Engineering and Surveying in 1878 in the Faculty of Science, was due to the prompting of a group of Bristol and Bath engineering firms and the City's Executive Committee. Its first two Professors were J. F. Main and Silvanus P. Thompson. Main, then Lecturer in Mathematics became Professor of Mathematics and Engineering, whilst Thompson, then Lecturer in Experimental Physics, assumed the title of Professor in that subject, plus those of Geometrical Drawing and Surveying.

It will be recalled that at this date the Society had not yet taken over the Bristol Trade and Mining School (they did so in 1885), so that relevant education in Bristol was largely concerned with teaching established practices. In contrast, the first Principal of UCB, Alfred Marshall, submitted a scheme to its Council in March 1878 which resolved to base its teaching on the scientific principles of engineering.

In developing this attitude, it is important to note that Marshall had been able to recruit as members of the College Council three people of like mind who would support his ambitions. Two of them, William Froude[1] and Henry Smith, were both Fellows of the Royal Society, and a third member, James Stuart, was Professor of Mechanisms and Applied Mechanics at the University of Cambridge. Its chairman was the same Gilbert Elliot who had made the 'Technical College of Science' proposal in 1873.

It is of passing interest to note that the Board of Governors of UCB was made up of three classes; donors who had contributed either £5 per annum or £50 for life, corporate donors who had contributed £10 per

---

[1] William Froude (1810–79), a Devon man who developed modelling techniques for experimental hydromechanics, particularly flow past submerged bodies; in doing so created the 'Froude Number' which we still use.

annum or £250 for life, and nominated officials and representatives. The voting system of this board would not appeal in our more democratic age, since it consisted of a sliding scale, the individual member receiving votes in proportion to their financial contribution.

Swift action was taken on Alfred Marshall's ideas of the principles to be adopted for engineering education because the College Calendar for 1878–9 contains the following description of the Department of Engineering:

> The instruction in this Department is designed to afford a thorough scientific education for students intending to become Mechanical or Civil Engineers, Surveyors or Architects.[2] The course for Engineering is such that students can persue it during the six winter months of each year, and the Council of the College have arranged with the following Civil and Manufacturing Engineers to receive into their offices and workshops during the summer months students whose position relative to firms would be that of articled pupils.

There followed a list of eleven well-known local firms from which further information could be obtained, including that of the fee which the student would be required to pay as an articled pupil, estimated to be between £30 and £50 which from its magnitude would have been for the whole course period of three years. From the above quotation it is clear that UCB intended to make a sharp distinction between education and training, their responsibility being for the former, whilst, although they committed themselves to organising it, training was the task to be undertaken by industry. We might now refer to such an arrangement as a 'thin sandwich' scheme of professional engineering formation, but with the great difference that the student paid the training firm, rather than the reverse arrangement which obtains today.

Courses were given in the Park Row building which was the first home of the UCB, and the course fee consisted of an entrance fee of 7 shillings for each course and 3 guineas annually for each of three years.

---

[2] Electrical Engineering did not feature as a distinct discipline, but Electrotechnology was taught by Prof. A. P. Chattock as part of Physics.

Evening courses were also given, each having two hours of instruction per week at a cost of 15 shillings for each year of three terms. Some of these evening classes were of an extra-mural nature in support of local industry; as, for example, lectures in chemistry and textiles for the woollen manufacturing activities at Stroud in Gloucestershire, Trowbridge in Wiltshire and Frome in Somerset. For the 1878 session there were 132 students in the day classes and 258 in evening classes. With the City of Bristol's record of international trade and enterprise, it is not surprising that a special booklet was issued in 1890 carrying the title 'Information for Colonial Students' in which the costs of a three-year course were estimated at £330 16s, to include '. . . lodgings, pocket money, books and instruments and passage money'. There were no halls of residence or other UBC-sponsored accommodation, so that the Calendar for 1878, and for many years to follow, contains the statement 'Professor Main is prepared to take charge of students whose parents reside at a distance.' Presumably his task was to find lodgings for these students.

In the first year of the three-year degree, the Calendar for 1878 shows that of the five separate courses, Main was responsible for the two vested in his title, and similarly for Thompson; in addition, Prof. Letts was called in to teach chemistry. In the second year, Main had machine drawing added to his responsibilities, whilst those of Thompson remained the same, with Letts switching to metallurgy. However, Main was totally responsible in the third-year, with mechanics added to his three second-year courses.

It would be interesting to know whether the actual teaching method adapted at UCB followed the German university pattern, in which the Professor gave all the lectures and his junior colleagues then took over the teaching role and were paid for doing so directly by the Professor. There is considerable evidence that this was the case, because when referring to Professors' salaries, the Departmental minutes state a figure (e.g. £300 p.a.) plus 'a fee for lectures given'. In further support of this contention, the minutes for November 1878 states 'Professor Main was permitted to engage Hele-Shaw[3] as an assistant in Mathematics and Engineering upon the understanding that the College is put to no

---

[3] Hele-Shaw (1854–1941). His Royal Society biographical memoir describes him as an inventive genius, chiefly remembered for his work in fluid dynamics, where he showed experimentally the existence of stream-lines in a viscous fluid.

expense.' It is also recorded that Thompson had an assistant, C. C. Starling, paid by himself. Within six months, reference is made to the official appointment of Henry Hele-Shaw to the post of Lecturer in Mathematics, Mechanics and Engineering, so long as 'Professor Main relinquishes fees of up to £100'.

Whether professors or lecturers were allowed, or even encouraged, to receive fees for the professional assistance which they gave to local industry is not known. Such assistance may have been part of their conditions of employment in view of UCB's continuing reliance on local financial support.

At the end of the 1882 session Professor Main resigned to take a post in London at the Royal School of Mines and what was then called the Normal College of Science, later of course to become the Royal College of Science, and in 1907, with the City and Guilds of London Institute, Imperial College. Main was replaced by Edward Buck as Lecturer in Mathematics, and Hele-Shaw (Fig. 2.1) as Professor of Engineering at a salary of £250 plus one-quarter of fees for all classes taught by himself, with a total minimum of £300 p.a. The inclusion of the phrase 'taught by himself' could be a clarification of the payment issue raised above, implying perhaps that junior academic staff would henceforth be paid directly from UCB central funds.

Because of the great interest in engineering education generated nationally by the 1851 Great Exhibition, many institutions, such as UCB, had been created in other parts of the UK, a consequence of which was a significant demand for scientific and engineering academic staff and much movement of such staff between these institutions. In 1885 both Hele-Shaw and Thompson resigned, the former to become Professor at Liverpool and the latter Principal of Finsbury Technical College in London. In the same year J. Ryan was appointed Professor of Experimental Physics and Engineering to replace Hele-Shaw, and in 1889 D. G. Selman (a former UCB student), who had held a lectureship since 1884, was promoted to a Professorship in Mathematics, apparently on the strength of his excellence in teaching that subject.[4]

By 1893 an increase in available accommodation had permitted separation of Engineering from Physics, which allowed Ryan to be re-styled

---

[4]   The idea of awarding personal professorship on the basis of teaching excellence, rather than research excellence, was revived in Bristol in 2005 when Dr. R. R. Clements (Eng.Maths.) was so rewarded.

Fig. 2.1 Two of the occupants of the Engineering Chair at UBC – Hele-Shaw to the left and Stanton to the right. Photographs of the other three – Main, Ryan and Silvanus P. Thompson could not be found.

Professor of Engineering and the appointment of two new lecturers. When Ryan resigned in 1899 he was replaced by Dr. T. E. Stanton (Fig. 2.1) – a junior colleague of Hele-Shaw at Liverpool – who later had a long career (1901–30) as Superintendent of the Engineering Department at the National Physics Laboratory, was knighted, and became a Fellow of the Royal Society. The probable reason for his short stay at UCB will be discussed later. He resigned in 1901 and was replaced as Professor by R. M. Ferrier, who, with F. J. Broadbent and J. Lees as lecturers, remained with UCB until its Department of Engineering – then part of the Science Faculty at UCB – became the Faculty of Engineering in the University of Bristol in 1909.

The Engineering section of the Science Faculty Prospectus for 1901 is noteworthy because, for the first time, it contains a special section on 'Information for Women Students', and it was repeated the following year. Of course, through the efforts of the first Principal's wife, an early Cambridge graduate and part-time member of academic staff, UCB had welcomed women students into other Departments from the very beginning, but this appears to have been the first occasion that Engineering was advertised as being a suitable career for women.

Fig. 2.2 Staff and students of the Faculty sometime between 1919 and 1925. Wertheimer (Dean) is in the centre of the seated row; on his left are, in turn, David Robertson, William Morgan and Edmund Bolton. On Wertheimer's right are, in turn, Robert Ferrier and Andrew Robertson.

Unfortunately, it is not known how many women students were recruited at this time. Even 20 years later, only one woman – standing in the back row – can be identified in Fig. 2.2, which is a photograph of staff and students, sometime between 1919 and 1925, in what had by then become the Faculty of Engineering in the University.

The course fees for 1876 were mentioned earlier; they were revised 20 years later to a composite fee of 50 guineas for the three-year course in Mechanical, 53 guineas in Civil and 80 guineas in Electrical Engineering; the reason for these differences was not given. For, comparison, a lecturer in Civil Engineering was appointed in 1898 at an annual salary of £100.

Having given a brief summary of the overall development of UCB, particularly with regard to its teaching programme, attention will now be concentrated on four of its constituent parts, research, accommodation, experimental facilities and student affairs.

2.3 Research

The attitude to research at UCB in its early years, in Engineering at least, can be gleaned from a request in 1889 by Professor Ryan. He asked

> to be released from his promise not to devote any time to original research, as he wished to undertake, with the assistance of advanced students, investigations which would not interfere with the performance of his College Engagements.

It would appear from this request that on his appointment, four years earlier, he had actually been requested not to engage in original research, but he must have sensed a change of attitude of the College authorities because at the next meeting his request was granted with the following proviso, 'on the understanding that scientific research only should be undertaken'.

As commented earlier, it would be surprising if Bristol industries had not wished to avail themselves of the expertise of the UCB academic staff for the solution of problems which were beyond their own capabilities, so was this at the centre of Ryan's request? If not, how did the UCB authorities interpret 'scientific research' and what, at that time, was their attitude to consulting activities? Whatever it was, it had certainly changed by 1905 because in the prospectus for that session the following sentence appears 'A considerable amount of research, most of it of *permanent value* (my italics) has been carried out,' and the following session's Calendar records 'The College provides University Teaching and affords facilities for research in all the important branches of Arts, Science and Medicine.'

2.4 Accommodation

The entire UCB was originally located in a long-demolished house in Park Row[5], but in 1876 a plot of land was purchased in Tyndall's Park, off what is now University Road. The Gothic-style proposal of the

---

[5] On this site a block of flats now stands, one of which caused a minor political scandal in 2005 when it was bought in a somewhat dubious manner by the Prime Minister's wife, Cherie Blair, for one of their sons who was a Bristol University student.

architect Charles Hansom was accepted because it envisaged a quad-rangle of four buildings which could be built in stages, the UCB not being financially able at that time to contemplate immediate construc-tion of all four buildings. Engineering was not re-housed in what became known as the 'Hansom Quadrangle'[6] – although the fourth side was never built – until the second stage, the west wing was opened in 1883 (Fig. 2.3). But it became clear immediately that its joint occupa-tion with chemistry was inhibiting the development of both Physics and Engineering, and it was not until 1893, following further develop-ment of the east wing of the Hansom Quadrangle, that Engineering was able to have its own accommodation, which for the first time provided a drawing office and a workshop. It was this new accommodation which allowed Engineering to increase the number of students to 60 instead of the 40 limit set in 1891, and to be separated from Physics. It was to stay there until it moved to Unity Street as the University Faculty of Engineering within the Merchant Venturers Technical College in 1909.

2.5 Development of Experimental Facilities

Despite the fact that early Prospectuses state that 'Special attention is given to class teaching and laboratory work in all courses,' no details were specified as to what the laboratory work actually was; whether it was 'hands-on', or merely group demonstrations by the teachers. It is clear that funds for both equipment and laboratory accommodation was sparse, and in 1878 we find Professor Main was given £30 'to obtain apparatus to illustrate his lectures on experimental mechanics,' and Professor Thompson was authorised to spend £35 on surveying instru-ments. In 1882 he complained about the shortage of apparatus and was given £257 against a request for £685. The first materials testing machine was obtained in 1884 by Hele-Shaw at a cost of £500, his clinching argument being that an income to the college of £40–50 p.a. would be obtained from local industry! Here is part answer to an earlier question in this history concerning the recipient of payments from industry for work carried out on their behalf by UCB staff; one recipient was clearly the UCB, but whether academic staff who had done the work received anything is not clear.

---

[6] Still in use in University Road; the Fry Tower was built later.

Fig. 2.3 Hansom's Quadrangle in University Rd., the E. Wing of which was occupied by the UCB Dept. of Engineering. It currently houses Geographical Sciences.

There was a steady accretion of equipment during the second half of this decade; in 1885, for example, £602 were spent on two further testing machines, a gas engine and tools for the workshop. But the February 1890 Minutes of the Engineering, Physical (*sic*) and Mathematics Sub-Committee record the full text of a letter from one of its members, the new Professor, T. E. Stanton, which drew attention to the 'lack of facilities for experimental work', citing the need for '... a steady supply of water to allow experiments on friction in pipes, flow over notches and weirs ...' in other words, he wanted to create a hydraulics laboratory! Although he was granted £126 18s 6d, Stanton resigned in the summer of 1901, but we do not know whether this was caused by the failure to get the facilities he wanted, or the attractiveness of a senior position at the National Physical Laboratory. There could have been another reason for his resignation, that of the routine nature of some of his responsibilities as is shown by the following three typical abstracts from the Minutes of the Sub-Committee referred to above:

> that the wood-working of the electrical lighting installation is
> allowed to remain unvarnished, that the desirability of renting
> a telephone was generally recognised, that the wages of
> Herbert Sweet be increased to 7s 6d per week.

Today's academic staff who have served on university committees may comment that nothing has changed; agenda items on which everyone has a view are likely to last a long time – often referred to as 'the bicycle-shed/nuclear reactor syndrome'. The present author is often given to recalling that one of his major achievements as Pro-Vice-Chancellor was, after more than an hour's discussion – with everyone contributing – taking upon himself a chairman's decision about the colour and style of the doorknobs on the new (1983) safety doors in the Wills Memorial Building. They are still there for all to see!

2.6 Student Affairs

By 1884 the success and value of the Engineering Department had begun to be recognised by the College authorities because in that year the Council established an engineering scholarship which it held to be justified '... In view of the present position of the Engineering Department and its direct influence in attracting students who attend courses of instruction *in other departments*' (my italics). The evidence for the last part of this quotation is not given, but it was probably some reflection on the personal characteristics of most engineering students, which continues to this day; namely – and as a group – the personal maturity which comes from a commitment to a professional discipline, and the significant contribution which they then made to the social, athletic and cultural life of the College. That they could combine a serious interest in all three of these activities was due to a lesser commitment to their academic studies than is necessary today. We have, for instance, the testimony of Prof. Andrew Robertson (Chapter 4) that in his student days at Owen's College, Manchester in 1902, 'Three engineering lectures a week sufficed, together with some lectures on mathematics and physics, and with "recommended" periods in the laboratory and drawing office.' As a result, whilst obtaining first-class honours in his degree, he was able to act as Secretary and Treasurer of the Engineering Society!

Moving forward to the early years of the Faculty, in 1919 the Dean was able to report that

> The social life of the University has developed greatly during
> the session, and in this development one of our students, Mr.
> F. H. Bullock, who was President of the Guild of
> Undergraduates, took a very large share; more particularly he
> spent much time in assisting in the foundation of the
> University Union, a Students Club which has its headquarters
> in the Royal Fort,

and in the next session, 'Three of the last four Presidents of the Guild of Undergraduates have been selected from the Faculty.'

The increase in tempo from this position, can be attributed in part to the creation of the University Grants Committee in 1919, which attached to an increase in government financial support an involvement in the detailed management of the university system, an involvement which, as we shall see as this history of the Faculty unfolds, increased in intensity as the years passed.

As the Faculty grew within an expanding University, the importance of the Students' Union was never questioned, and throughout its existence Faculty academic staff took a part in assisting with its management. With this insight, concern grew that the Union Officers were increasingly being elected from the 'non-laboratory' faculties, simply because students in Science, Medicine and Engineering had academic schedules which occupied the greater part of their time. In 1962 the Faculty expressed their concern in Senate, presenting the data in

Table 2.1 Faculty Membership of the Union Council, 1957–62

	1957	1958	1959	1960	1961	1962
Arts	8	11	8	10	9	9
Science	3	2	6	4		
Medicine	2	1	1			
Law	1	1	1	1	5	6
Engineering	1					

Table 2.1, but few members were seriously concerned, accepting the anodyne resolution:

> . . . to call the attention of the Union to the question of faculty representation amongst its officers and to invite it to consider how the Union Council might be enabled to function satisfactorily in the enlarged University of the future.

Its response a few years later, was to successfully argue that because their duties would expand as the University grew, its six principal officers should be allowed to take a sabbatical-year during their term of office, making it even more difficult for students from the 'laboratory' faculties to stand for election. It was to become virtually impossible after 1997 when two-year sabbaticals were proposed.

Chapter 3

Creation of the University of Bristol Faculty of Engineering

The Agreement between the MVTC and UCB, 1901–09

3.1 Introduction

We have already seen that in the last quarter of the nineteenth century there were two main streams of what we can properly call engineering education in Bristol, one provided by the MVTC as part of its activities, and one at UCB; but there were others which provided technical training, notably the Bristol School of Practical Art which was founded in 1853 by the government department of the same name to provide 'Artisan classes in the evening and classes for Ladies and Gentlemen in the day-time.' These classes included geometry, building construction and elementary architecture, presumable for the artisans, and various types of drawing for the ladies and gentlemen. It was also the stated intention of the Bristol Technical Instruction Committee of the City Council to open a school of science in two different parts of the city, but these had no bearing on our history.

During the last decade of the nineteenth century the demand for education grew at an increasing rate, the MVTC for example having a total of day and evening students of 1430 in 1890, but 2215 in 1902, with adult day students making up only 279 in this number, so that much the larger body was of what the literature of the time refers to as 'artisans'.

The overcrowding of their accommodation in Unity Street was not the only concern of the governors of the MVTC at this time; they were also concerned with what they considered to be overlapping and competition in the provision of their higher level courses, the prime competitor in their view being UCB. In an attempt to assuage this concern, the Governors sought the assistance of the Government Department of Science and Art, who referred it back to local level, with the result that the MVTC and the School of Practical Art agreed areas of activity which allowed the MVTC to concentrate more on advanced work in science and engineering. One result of this change was that by 1891 it was offering courses leading to the BSc degree of the University of London.

3.2 Involvement by the City Council

Whilst the foregoing issue was being considered by two of the participants, the City of Bristol itself became involved through a letter from the Town Clerk to the governors of the MVTC pointing out that there was considerable overlapping in the classes provided by the MVTC and the UCB. This letter contained the following suggestion:

> ... that the two colleges should get together in an attempt to do away with the overlap. Each college would then be able to develop to its fullest extent – perhaps even to combine into one institution, with one governing body – and the Technical Instruction Committee might then be able to give much material help to such an institution. The Technical Instruction Committee have reason to believe the Government would give a much larger subsidy than that now enjoyed by University College.

The implications of this letter are very clear; if you want further financial support from tax and rate payers, settle your differences!

It will be recalled that the Society of Merchant Venturers were actually sympathetic in 1876 to the founding of UCB and had contributed £1000 to its appeal fund, but by 1896 it was less sympathetic, and had declined an appeal for a further contribution. The dispute which now appeared between the two bodies centred on the UCB view that the MVTC's original aim was 'to provide a complete, continuous and thoroughly sound preparation for an industrial career', and that by developing the depth of its courses to BSc level, the MVTC had encroached upon the very ground which Alfred Marshall, the first Principal of UCB, had staked out, that of '... teaching which the College gives in the scientific principles of engineering'. The MVTC refuted this, arguing that since 1885, and more particularly since 1894 when it changed from School to College, it had been teaching the sciences related to engineering, mining, metallurgy, manufacture and commerce. It also pointed to the difference in course fee levels, the higher ones at UCB making it impossible for the poorer students – the artisans – to attend. The governors of the MVTC supported this contention of both utility and success by pointing out, in relation to utility, that in the 1898 session they had had a combined total of day and evening students of 1426, whereas the whole Arts and Science Departments at UCB had only 348 students in comparison. Relating to success, 12 students at the MVTC had obtained the BSc degree externally from the University of London, but only three had done so from UCB.

An early positive suggestion came in 1901 from the MVTC. It set out a sub-division of subjects into three categories, one each to be the preserve of either the MVTC or UBC exclusively, and a small group of subjects containing pure mathematics, pure chemistry and pure physics which would be common to both; what the adjectival use of the word 'pure' implied we do not know. In essence, the UCB subjects were to be in Arts and Medicine, whilst those of the MVTC included '. . . all technical subjects including Engineering'. In conclusion, the MVTC suggestion proposed that if the two colleges were to be merged, it should be called 'The West of England University and Technical College'. In using the word 'merger' it became clear later that the MVTC meant a union of partners having different governing bodies, citing the University of London as an example, whereas UCB wanted a complete fusion of the two colleges, having one governing body. When negotiations broke down, a senior member of the Society attempted to bring the two colleges together again and a meeting was arranged between three representatives of each side, G. H. Pope, W. W. Ward and J. Wertheimer for the MVTC, and Albert Fry, J. W. Arrowsmith and C. Lloyd Morgan for UCB. Although they agreed

> . . . that the object of the conference is to secure for Bristol the best possible scheme of higher education and incidentally as a means to an end an increased Treasury grant.

the main disagreement – on the role of Engineering in each college – remained, causing formal discussions to be discontinued, whereupon

UCB proposed to submit its own ideas in a petition to the Privy Council for a charter establishing it as the University of Bristol.

3.3 Political, Religious and Social Differences

These conflicts of interest in educational matters were certainly not improved by basic political, religious and social differences, as well as differences of temperament between the principal negotiators. Those who created UCB, including the Wills and Fry families, were Liberals in politics and non-conformists in religion, whilst members of the Society were generally Conservatives and members of the Established Church. It is a reflection on latter-day social mores, that the Wills and the Frys, having made large and recent fortunes from tobacco and chocolate, respectively, were *nouveau riche* and were not members of the Society of Merchant Venturers. Of the leading negotiators for the MVTC, Julius Wertheimer, having himself raised his college from a school which included primary, secondary and basic training components, to a technical college of national repute, was not only jealous of its reputation, but also of his position as its Principal. He was clearly an excellent, though aggressive administrator, but with no pretensions of achievements in scientific or engineering research. In contrast, Conwy Lloyd Morgan, Principal of UCB since 1891 and formerly its Dean and academic head, had a major reputation for research in geology and zoology, and was later to enhance it further in psychology and ethics. It is less than surprising therefore, that Wertheimer the administrator, and Lloyd Morgan the academic, could not agree on the detailed nature of the proposed university.

Negotiations had finally broken down in 1901, but in the same year the MVTC lost part of its jealously guarded autonomy by accepting £3000 p.a. from the City's Technical Instruction Committee, which not only required the MVTC to accept two members of that committee on to its governing body, but also that it make further attempts to reach agreement with UCB about overlaps on the provision of engineering education. A second matter which, as we shall see, had an important bearing on an agreement with UCB, was the complete destruction by fire in October 1906 of the MVTC's building in Unity Street. Fig. 3.1 gives an artist's impression of the exterior view before, and after, the fire, and Fig. 3.2 is a similar impression of its attractive interior.

MERCHANT VENTURERS' TECHNICAL COLLEGE

MERCHANT VENTURERS' TECHNICAL COLLEGE

(Design adopted for the rebuilding scheme after the fire of 1906.)

Fig. 3.1 An exterior view of the Merchant Venturers Technical College before (upper), and after (lower), the fire in Oct. 1906.

Meanwhile, in the City of Bristol general support for UCB was rising, for in 1899 the University College Colston Society was established with J. W. Arrowsmith as its secretary, and Napier Abbott, a Bristol lawyer who acted as solicitor to the Wills family, as a prominent

The Great Hall

HALL OF THE DESTROYED COLLEGE

(The large hall of the Merchant Venturers' Technical was remarkable for its carved wood ceiling and panelled walls.)

Fig. 3.2 The interior of the Merchant Venturers Technical College before the fire in October 1906.

member. Its prime purpose was to raise funds to help the college and to generate support for a West of England University, to include not only UCB and MVTC, but colleges in Reading, Southampton and Exeter as well. Amongst its supporters outside Bristol in 1901–02 were John Percival – one of the creators of UCB – formerly Headmaster of Clifton College but now Bishop of Hereford, Augustin Birrell, one of the Bristol members of Parliament and President of the Board of Education at the time, and R. B. (later Viscount) Haldane. Much of the correspondence between these five (and some others) has been preserved, from which it is clear that Arrowsmith was the pivotal figure, having an easy personal rapport with the other four. Jocularity in letter writing was not usual at the beginning of the twentieth century, but we find Abbott addressing Arrowsmith as either 'Coppersmith', 'Arrow' or 'Sagittarius', all three appellations being a play on Arrowsmith's name.[1] Although 'Abbott' might have given Arrowsmith the opportunity for responding in like manner he did not do so, although his letters were still very informal in style.

As a Bishop, Percival, sometimes writing from Lambeth Palace, was a peer of the realm and at the centre of affairs in London, as was Burrell of course, and the correspondence with Arrowsmith showed that these two, together with Haldane, guided him and the other UCB supporters in their eventual dealings with the Privy Council, which will be described later. But in the matter of engineering education, which, we remind ourselves, was the basic cause of friction between the MVTC and UCB, the support of Haldane was likely to have had the greater significance. Not only was he a lawyer and member of the Liberal Party, eventually becoming Lord Chancellor in the Liberal Government of H. H. Asquith (1908–15), but by religion a humanist rather than a member of the Established Church, characteristics which would have inclined him to be sympathetic to those supporting UCB rather than to the MVTC sponsored by the Society. Of even greater importance, his attendance at various universities included a period at the Technische Hochschule at Charlottenburg in Germany, where he observed science and engineering education being carried on within an environment which included languages, literature and the social sciences. When he acquired authority as a government minister, these experiences caused him to make great efforts to promote scientific and engineering education in Great Britain in a similar style. In his address at the third Annual Dinner of the Colston Society in 1902, he told the audience that it should look to Germany as a model for national prosperity, through education which combined the theoretical with the practical, suggesting that Bristol itself should look at cities such as Leeds, Liverpool,

---

[1] Sagittarius, the archer – a sign of the Zodiac.

Manchester and Birmingham as examples of what his activities had already achieved, and which Bristol should follow. In a personal letter to Arrowsmith he pointed to Joseph Chamberlain as the key financial catalyst in Birmingham, suggesting, without names – but obviously with the Wills family in mind – that Bristol should look for a similar sponsor. In another letter, written in response to a question from Arrowsmith, Haldane tells him that the Privy Council would expect financial backing of £100,000 before it would consider a proposal for a Royal Charter, and that support from the City of Bristol would also be required, ideally in the form of a grant from its rate income. Haldane was not a scientist himself, but his influence and success in such endeavours earned him Fellowship of the Royal Society in 1906. At Liverpool he was influential in creating the university there, and in forming Imperial College, London in 1907, by merging the Royal College of Science, the Royal School of Mines and the City and Guilds College.

By March 1906 £30,000 had been raised by UCB, most of it from the Wills and Fry families, and possibly as a result of letters in March 1906 from both Haldane and Bishop Percival urging Arrowsmith to ignore the opposition of the MVTC and to press ahead, a committee to create a university was formally established. The developments of pressure from the City, the possibility of a federal type of university which it had proposed earlier and the loss due to fire of its building, forced the MVTC to enter once again into discussions with UCB. There was, however, no substantial change of attitude by either side. The MVTC still wanted a federal university, whereas UCB wanted a unitary one on a single site. Talks which took place over two years made no progress, which was not surprising since the same people, maintaining the same positions, were involved. The MVTC took an uncompromising stance, insisting that engineering and technical education at all levels should be its sole responsibility. But, feeling that it was negotiating from a position of strength now that its rebuilt premises in Unity Street were in use, it appeared to overplay its hand, even to the extent of demanding that if a new Faculty of Commerce and Economics should be established in the future, it would be its responsibility, not that of UCB.

3.4 The Crucial Intervention of H. O. Wills

The tone of the negotiations took a decided turn in favour of UCB when in February 1908, at the University College Colston Society's dinner, it was announced that Henry Overton Wills had promised £100,000 towards the endowment fund necessary to create a University of Bristol on condition that a charter was obtained within two years. This amount was one-half of the £200,000 which had been estimated as necessary two years earlier, but was exactly the amount which Haldane had said the Privy Council would require any organisation to possess before it would consider granting it a charter. With existing endowments the total then raised was around £150,000 and the UCB realised that although it would be valuable to have the MVTC in collaboration, it was no longer essential. The cartoon in Fig. 3.3 makes it reasonably clear that Viscount

Fig. 3.3 A cartoon showing H. O. Wills delivering £100,000 to J. W. Arrowsmith and Napier Abbott (suitably dressed!), with J. B. Haldane and Augustin Birrell in the doorway.

Haldane and Augustine Birrell had been influential in persuading their fellow Liberal, H. O. Wills, to take a serious interest in a university for Bristol and to make this donation. The cartoon shows the approach of Henry Overton Wills with a bag containing £100,000, and J. W. Arrowsmith the secretary of the Colston Society whispering in the ear of an appropriately dressed Napier Abbott, a keen supporter of UCB and later the university. In the doorway are Augustin Birrell, and the round face of Haldane, who later received recognition for his influence by becoming the second Chancellor of the University in 1912.

From here on, UCB elaborated its views and conditions that there should be complete unification of the work of the university under one administration, and that degree-level science and applied science in both MVTC and UBC would be completely merged in the university; some would be found in the Faculty of Science, but the greater part would be in a Faculty of Applied Science and Engineering with which the name of the Society of Merchant Venturers would be associated. Further, the second of these two faculties would occupy the MVTC building in Unity Street. To deal with other levels of activity in the MVTC, the UCB suggested that for 'the technological instruction of artisans' a Merchant Venturers School of Technology should be created, to be managed by the University, the Society and the City Council.

3.5 Charter Applications by UCB and MVTC

As would be expected from its previous stance, the MVTC did not accept the UCB's scenario of the future of engineering education in Bristol and reported so in April 1908, whereupon the UCB informed them that they were, nevertheless, intending to apply to the Privy Council for a charter. In doing so, they informed the MVTC that copies of the correspondence between them on this issue were to be sent to the City authorities. With the issue now in the public domain, the MVTC published its own account in a pamphlet 'The Society of Merchant Venturers and the proposed University of Bristol and the West of England.' It contained all the previous claims, including that it was first in the field of engineering education in Bristol and that the idea of a federal university with the MVTC as a self-governing part, had been proposed by the Bishop of Bristol with support from other notables such as R. B. Haldane and Bishop Percival. On financial aspects, it said that the financial strength of the Society had been wrongly assumed, by both UCB and the City authorities, and that, in fact, it was unable to support both a Faculty of Applied Science and Engineering in the University and a separate Merchant Venturers School of Technology.

The City again intervened by suggesting a meeting between the two sides, also reiterating that it would not countenance duplicate grants in the future. The meeting took place, but the determined and opposing views of the people involved did not allow a solution to emerge, whereupon the UCB proceeded with the preparation of its petition to the Privy Council for a charter. It had the opportunity to further its cause by presenting a Loyal Address to King Edward VII when he visited Bristol in July 1908, but the Society countered in a separate Loyal Address which, although specifying a willingness to undertake the provision of a Faculty of Engineering of the University, did not specify whether the university should be federal or unitary. When both the University committee and the Society presented petitions to the Privy Council a few months later, the Society made it quite clear that it had not moved from the position of insisting on a federal arrangement between the two colleges, the MVTC retaining complete financial and administrative control over the provision of degree-level courses in engineering, ceding only academic matters to the new University.

When the draft charter for the creation of the University was presented to Parliament in November 1908, it did not incorporate the MVTC. It will be readily understood that this caused such concern as to generate immediate contact between the Society and the Privy Council, with allegations of bias by the latter against the former, coupled with suggestions of private dealings between the Privy Council and the University Committee. Political influence was also raised as a possibility, based upon the fact, previously alluded to, that the University Committee had Liberal tendencies[2] and were therefore supporters of the Liberal government then in power.

Details of the strenuous activities of the Society's representatives on behalf of the MVTC in their protestations to the Privy Council are given an excellent treatment by Prof. McGrath (Ref. 1) and will not be repeated here. It is sufficient to say that their persistence was such as to cause the Principal Secretary of that Council to lose his customary

[2] The relationship between the Liberal Party, the Wills family and W. S. Churchill, a future Chancellor of Bristol University (1929–1965) will be mentioned later.

administrative coolness when writing to the Society in the following terms:

> After a very considerable experience of negotiations in
> connection with University Charters, I am free to confess that
> I have never had to deal with a body of gentlemen whose
> disposition was so fluctuating and mistrustful. You have been
> met in a way which amounts to a very handsome recognition
> of your special claims and all you do is to raise fresh points
> and extend your pretensions to matters lying outside the
> purpose of the agreement.

From this it may be construed that the Secretary's previous experience had not included dealing with such a prestigious and powerful body as the Society of Merchant Venturers of Bristol, who were clearly fully supportive of their negotiators.

3.6 The University Receives its Charter

But there were indeed many important issues of detail which still had to be resolved. Amongst these was whether the new Engineering Faculty should teach chemical engineering, the alternative being the Faculty of Pure Science, and the conditions of employment of the staff of UCB who were to be transferred to the Faculty of Engineering at the MVTC building in Unity Street. Wertheimer pointed out that their teaching load was relatively light compared to that of the MVTC staff, and his opinion of Professor Ferrier – the Head of Engineering at UCB – particularly as a researcher, was expressed in his usual astringent terms. In any modern assessment of their relative prowess in research this would be considered as a case of the 'kettle calling the pot black'. It is true that Ferrier had almost no publication record, but Wertheimer's, certainly while he was Principal of MVTC, were on matters of educational administration rather than engineering science. Notwithstanding this personal animosity, of even greater importance for resolution between the two parties were precise details of the sources of finance for the new Faculty of Engineering.

At this stage there must have been some weariness in the efforts of the Society's negotiators, but it is also probable that its wiser members began to see the advantages in being associated with a multi-faculty

Fig. 3.4 The first six Faculty of Engineering Professors; in order from 1 to 6 they are, Julius Wertheimer, David Robertson, Edmund Boulton, John Munro, Robert Ferrier and William Morgan.

university rather than a technical college, together with the greatly increased financial contribution from government and the City of Bristol, where the corporation had in April 1908 voted a penny rate (*c.*£7000 in 1909) to the projected university after its town clerk had taken advice on the level required from his colleagues in Leeds, Liverpool, Manchester, Sheffield and Birmingham. It was certainly necessary to mollify Professor Wertheimer, and this was achieved by making him Permanent Dean of the proposed new Faculty of Engineering (with Ferrier as a subordinate colleague), a post which he was to hold until his death in 1924. Whether at this time he was promised an honorary degree is not known, but along with a large number of other participants in this saga, he was so honoured by the award of a DSc in the 1911 degree ceremonies.

In May 1909 the Society approved the terms of agreement sent to it by the Privy Council office, and later that month, actually on the 17 May, it was informed that the King had approved the charter.

Later in the same year the Society put its seal on the formal agreement with what was now the University of Bristol, to provide a Faculty of Engineering within it, to be housed in the MVTC in Unity Street. The agreement, covering eleven A4 pages, naturally sets out clarification in legal language of the many contentious issues previously mentioned. Appendix 1 is the author's summary of its essential features; the numbering is not that of the agreement and explanatory comments have been added where appropriate.

The senior academic staff transferring to the Faculty from MVTC were actually appointed by the Society and then 'recognised' by the University. They were, Julius Wertheimer of course, Edmund Boulton (mathematics), John Munro (mechanical and mining), David Robertson (electrical), William Morgan (motor car) and Samuel Reynolds (geology), all of whom were accorded the title of Professor in the University; whether this was part of some informal element in the agreement is not known. Apart from Reynolds (his photograph has not survived), they are shown, together with Ferrier (from UCB), in Fig. 3.4. Other transferred staff are listed in Chapters 11–16.

Chapter 4

The First 40 Years – 1909–49

4.1 The New University of Bristol

The first meeting of the University Senate – made up of 21 professors from the four faculties of medicine, arts, science and engineering – was held on Friday, 9 July 1909 at 4.30 p.m., with the University Council meeting at the same time. Apart from Wertheimer, three of these Senate Professors, Boulton, Ferrier and D. Robertson, were from Engineering. At the Council meeting, the Vice-Chancellor, Conwy Lloyd Morgan, was accompanied by only one other member of Senate – Prof. Wertheimer, which was an indication of the status which the Principal of the MVTC had demanded and received. The remaining Senate members were required to wait in an ante-room for these two leaders to join them before they could start their own meeting!

There was so much to be done in creating a new university that between July 1909 and September 1910 Senate met twenty-three times, starting initially at 8.30 p.m. (*sic*), but soon bringing the time forward to 5.30 p.m. At the first of these meetings of Senate, Conwy Lloyd Morgan had announced his resignation as Vice-Chancellor with effect from 1 September 1909, allowing Sir Isambard Owen[1] to be elected in

---

[1] The interesting Christian name derives from the fact that his father was Isambard Kingdom Brunel's resident engineer on many projects, and IKB was Owen's godfather. The engineering connection is also recorded in the fact that as Vice-Chancellor, Owen suggested that the Wills Tower should be built in concrete, but the architect George Oatley (see also Chap. 5.2) said he wanted the structure to last as long as the Oxbridge colleges, and with an opinion that subsequently proved to be correct, declared that concrete would not last that long!

his place. In this appointment the University were extremely fortunate. Owen, whose formal qualifications were in medicine, had already been successful as a university administrator at Newcastle and in Wales; in the former by reconciling differences between the School of Medicine and Armstrong College, and in Wales he had himself written the charter creating the University of Wales from four previously independent colleges. His skill in managing university affairs involving people was quickly in evidence in Bristol. So as to make clear to Wertheimer that he, Owen, was the Vice-Chancellor and leader of the new University, one of the existing professors, Michelle Clarke, was appointed to the newly-created post of Pro-Vice-Chancellor, to work with Wertheimer in producing the detailed Ordinances and Regulations which were required. This was the kind of task which suited the administrative skills already shown by Wertheimer in raising the MVTC to its present position as a leading technical college, and the records show that it was he, using his greater experience, who initiated most of the necessary proposals for decision by Senate. One of these regulations is noteworthy in that it allowed study in another academic institution to be taken into account when awarding degrees. Its immediate significance was of course to existing students at the MVTC, but, as we shall see (Chap. 8.11), almost a century later it was to have benefit when UK universities introduced 'modularisation' of degrees and 'credit transfers' between universities.

As a result of a written request from the Guild of Undergraduates, Wertheimer and three other members of Senate were required to decide on the colours to be used for academic costume. How long they took over this task is not known, the eventual decision being scarlet, pale-blue, old-gold and silver-grey for medicine, arts, science and engineering, respectively. What is known, is their decision that hoods must be '... a colour called University red'.

In dividing subjects between the Science and Engineering faculties, mathematics, physics and chemistry were to be with Science, whilst applied chemistry, applied mathematics, and civil, mechanical, electrical and motor car engineering were with Engineering. The first subject in the second group may appear anomalous, but it gave the Dean his academic credentials in the new Faculty and it continued to exist until 1927. A second apparent anomaly was Geology, which became part of the Faculty with Prof. Sidney Reynolds as its Head, and it continued so until 1950 when it transferred to the science faculty. After that the Faculty Prospectus records that 'The Departments of Geology,

Economics and English also provide courses in the Faculty of Engineering.' As we shall see in Chapter 5.2, in 1950, despite being part of the Science Faculty, Geology was allocated a considerable volume of space in the Engineering Faculty's new building.

4.2 A Faculty within the MVTC

The agreement on the future of Engineering education in Bristol between the Society of Merchant Venturers and what had recently become the University of Bristol, was not signed until late in 1909 (Appendix 1), but preparations for its implementation had been made during that year, so that a Faculty Prospectus for the 1909 session was published. It is important to realise that this agreement was for the Society to provide and maintain the Faculty of Engineering of the University as part of the MVTC, with an amalgamation of both staff and students; it was not that all activities previously carried out by the MVTC should be taken over by the University. As a result, the MVTC continued to provide both day and evening classes, but as time went on such classes were more and more restricted to the technical aspects of engineering education, leaving the new Faculty to concentrate on the style of Engineering education which UCB had been promoting from 1876 onwards. It should also be noted that within a short time the total cost of maintaining the Faculty was not borne by the Society alone, but involved an increasing contribution from the University, which in reality of course came in greater part from government and Bristol City funds.

The first Prospectus sets out the courses to be given in both day and evening classes and the fees to be paid. For day classes there was a registration fee of one guinea (£1 1s) and 25 guineas per annum for a three-year course, reduced to 10 guineas if the parents earned less than £350 p.a. This figure appears to be high, perhaps deliberately so in order to attract students, because in a letter from the Registrar, Mr Rafter, to Prof. Wertheimer, the salary of Prof. Leonard (History) is quoted as £300 p.a. plus fees of £50, and in an advertisement for a lecturer in Civil Engineering in May 1916 the salary offered was £180 p.a. rising by increments of £10 to £200 p.a. For evening classes the registration fee was 5 shillings, and for attendance on three evenings per week the fee was 15 shillings for a course lasting one session. A fee of one penny was charged for each of the Saturday afternoon classes for miners. For the day students, the BSc in Engineering could be obtained in three years, but evening students were allowed to spread their studies over as many years as they wished, with a minimum of five. It appears that courses in both French and German had to be taken, the lecturer in French being Charles A. L. Dirac, the father of Nobel prize winner P. A. M. Dirac who in 1918–21 was a student in the electrical engineering department. The lecturer in German was C. C. Hennig, and these two continued until 1920 when Dorothy Shepherd provided the teaching in both languages until 1925. In 1927 Edith Bennett took over the role, providing her with employment until 1946, after which no Faculty appointment was listed for language teaching. Because the normal entry requirement after this date was the Ordinary and Higher School Certificate, with the former containing at least one European language, it must be assumed that the Faculty considered it to be unnecessary to provide formal language teaching.

As an indication of the organisational situation existing at that time, for the 1910 session the Annual Report was headed

Merchant Venturers Technical College
(*in which the Faculty of Engineering of the University of Bristol
is provided and maintained.*);

it was signed in the style 'J. Wertheimer, Principal'. Also noted was the fact that the session was the second of the Faculty but the fifty-fifth of the College. The MVTC had over two thousand students whereas the Faculty had only eighty-two, of whom 58 per cent had matriculated[2] compared to the national average of 22 per cent. Because this statistical measure rose to almost 100 per cent over the next two decades as an indication of the success of the Faculty in attracting well-qualified students, it should be recorded that matriculation examinations were entrance examinations sponsored by many universities at that time. There was also a University of Bristol School Certificate examination, established in 1912 for accredited schools in Bristol and the neighbouring counties. Even after the 1939–45 World War such examinations co-existed for some time with the national examinations of School Certificate and Higher School Certificate, which were themselves replaced later by the General Certificate of Secondary Education (GCSE 'O' and 'A' levels).

---

[2] Latin *matriculare* – to enlist or enrol.

Returning to the 1910 Annual Report, there are three points which are especially noteworthy. The first is that although most students entering the Faculty came from the southern counties of England, there were students from Ireland, India and South Africa, and later annual reports were to show an extending international catchment area. The second point of note is that of the 16 students graduating that year, Mr. A. J. S. Pippard not only carried off all the prizes, but was also successful in the national competition for an Industrial Bursary awarded by the Royal Commissioners for the Great Exhibition of 1851, which excused him from fees. Pippard subsequently became the second Professor of Civil Engineering in the Faculty, President of the Institution of Civil Engineers and a Fellow of the Royal Society for his work on strain-energy methods of structural analysis. But Pippard is not the only graduate whose name we should record from this cohort of students – Reginald E. Stradling is another. He was a Bristol Grammar School boy who was awarded the MC (Military Cross) for gallantry during the 1914–18 World War, before joining the University of Birmingham to carry out research on the properties of building materials, for which he eventually obtained the DSc degree from Bristol. Having moved into the Scientific Civil Service (DSIR) he became Director of Building Research and subsequently Chief Scientific Advisor to the Ministry of Works. During these appointments he created the Steel Structures Research Committee which, as we shall see, greatly influenced the academic careers of three of our Faculty Professors. Added to these achievements, he married Pippard's sister, was elected to the Royal Society in 1943, and received a knighthood in 1945.

The final point of significance in this Annual Report is that care was obviously taken in the choice of speaker for the annual prize-day address. He was Alexander Siemens, who, apart from creating the international firm of electrical engineers which bore his name, had the very rare distinction of being a Past-President of both the Institutions of Civil and of Electrical Engineers. For his address he took as his theme the Aristotelian definition of the difference between science and art – a definition which has been referred to earlier in relation to the creation of the government Department of Science and Art in 1852. In essence, Siemens pointed out that 'art' is to be interpreted as applied science, in which the output is influenced by the knowledge and ability of the producer, whereas in science itself, the conclusions are independent of the producer, being governed by immutable natural laws. On this issue he concluded his remarks by

> Does experience not suggest that a sound preparation at the College should be directed to the science side of the profession or trade which the student desires to follow, leaving the practical training of the student in other hands.

Those present, who had worked so hard to create the Faculty, must have been greatly pleased to hear these words spoken by such an eminent member of the engineering profession and a successful leader of industry.

In addition to the regulations for the BSc(Eng.) degree, those for the MSc and DSc were also set down, with much the same conditions as now, although there was an option to obtain the former by a minimum of two years in engineering practice plus a dissertation. Anticipating the University Appointments Board which appeared in 1923 and its successor the Careers Advisory Service which was to come much later (1970), the Faculty considered it to be a duty to find employment for its graduates as the following extract indicates:

> The following firms are willing, other things being equal, to give preference to students who have completed the course here and are prepared to take students into their works at *reduced premiums* (my italics).

A list of 12 local firms is given.

4.3 The Faculty Starts its own Life

By the 1911 session the strength of the Faculty relative to that of the MVTC had grown measurably, as is evidenced by the fact that there were two annual reports, one for the College and one for the Faculty, the former signed by Wertheimer as Principal of the College and the latter by him as Dean of the Faculty. As well as being accorded the title of Dean, a post which he retained until his death in 1924, Prof. Wertheimer had been given an honorary DSc by the University in 1911 and these advancements in his status might have had a bearing on the relative importance which he began to attach to his two rôles. They did not, however, smooth the roughness of his dealings with others. Existing correspondence between the two showed that he clearly believed himself to be at least the equal of the Vice-Chancellor, Sir Isambard Owen. There certainly were major issues to be clarified between them, such as the equalisation of salaries between the former MVTC and UCB staff, and whether the new Faculty should pay them termly, as at MVTC, or quarterly as at UCB. But correspondence between them shows that Wertheimer also pursued trivial issues, such as insisting on having a carpet in his room, his order of precedence in university processions, and whether all the books promised by UCB had been delivered. Another issue, which was not trivial, was the appointment of existing UCB staff to the new Faculty. Wertheimer felt that some were incompetent, and that their appointments should be terminated, but he succeeded in only one case, by a stratagem of a two-year period of what would now be referred to as 'gardening leave', with salary but no duties. It is clear that he continued to act like a Headmaster rather than a Dean, especially regarding the academic staff, as the following extract from an advertisement for a lecturer in civil engineering shows:

> The lecturer will work under the general direction of the Principal of the Merchant Venturers Technical College (who is also the Dean of the Faculty of Engineering) and under the immediate supervision of the Professor of Civil Engineering; he will not be allowed to take any outside work, unless with the sanction of the Principal given on the recommendation of the Professor.

It was not until May 1919, when Wertheimer submitted to Owen the draft of an advertisement for the Chair in Mechanical Engineering which contained the phrase 'The Professor will be directly responsible to the Society, which looks to the Principal of the MVTC . . . to be its representative,' that the Vice-Chancellor felt that he had to object formally to Wertheimer's 'personal rule' of the Faculty. With an astuteness which he had shown throughout his career, Owen told Wertheimer that the imposition of sanctions was a matter for the Society alone, and not one of its representatives!

The change of style and size of the annual Faculty Reports, which are now the only ones of concern to us, was accompanied by changes to its contents, particularly regarding research. In the two earlier MVTC reports a single paragraph of a few lines tells us that the amount of research was increasing, but no actual publications were reported. In contrast, the 1912 Faculty Report, of 12 pages, gives detailed account, department by department, of researchers and the work in progress, followed by a list of publications, a list dominated by the Electrical Engineering Professor, David Robertson, who authored nine of the 12 listed. Another new feature was a list of external examiners in civil, mechanical, electrical and motor car engineering, as well as chemistry, geology and mathematics. The fact that mathematics is still taught today in the Faculty by its own Department of Engineering Mathematics, rather than, as in most other universities by the university department of mathematics, owes much to the carry-over from this period. During its 100 years of existence there have been several attempts by the University to bring the Faculty into line on this issue, but resolute leaders have fought, and won, its retention, principally on three grounds: first, that 'service' teaching by one department to another is commonly regarded as a chore; second, that experience has shown that students are more likely to take a serious interest in mathematics when they are taught by academic staff who can use their own experience to show where it has been applied in practice; and third, that as the present author can testify, close contact between mathematicians and engineering academic staff does lead to very fruitful research.

Because this 1912 Faculty Report set a precedent which is followed for many sessions into the future, it is valuable to record three further innovations. They are, the creation of the University Engineering Society; the fact that the Faculty had offered its accommodation to Professional Engineering Institutions and Societies in Bristol as a venue for their meetings; and the inclusion of a list of projects which had been undertaken at the behest of local industry. The first President of the Engineering Society was the Professor of Automobile Engineering, W. Morgan, and it is clear from the fact that five students presented papers during the first session that it provided the valuable educational function of allowing students to present their work to an audience, and to cultivate the mental agility of quick thinking by answering questions from the audience. The Engineering Society also organised visits to six local industries – the inevitable brewery of course – but also to the Bristol Tramways and Carriage Co., later to become Bristol Aeroplane Co. and sponsor of our Chair in Aeronautical Engineering.

The offer of accommodation to the local branches of the Professional Engineering Institutions – which continues to this day – meant that both academic staff and students could become imbued with the professional ethos which these bodies generated. These Institutions were of course both learned societies and qualifying bodies, membership of which, though not mandatory, was of great value to students who wished to practice their profession at anything other than the lowest level. The fact that the Institutions had been encouraged to hold their meetings in the Faculty allowed students to meet senior engineers and to understand more readily what their choice of profession entailed. During the next 100 years these beginnings played a part in nine professors of the Faculty being elected to the national Presidency of these Institutions, and many more as Vice-Presidents.

The role of the Faculty in carrying out work for local industry appears from this 1912 Dean's Report to have consisted of tests on either materials or machines. There is no mention of payments for such tests, whether they were to the University, Faculty or to individuals, but this, and many succeeding reports, indicate that the Faculty benefited greatly from gifts of materials and machines from these firms. Because this work for industry was recorded in a separate section from research, we can only assume that industry did not sponsor basic research at this time. For example, the very active Prof. Morgan had four topics listed under 'research', and six under 'industrial work' but there were no links between the two sets, and none of his industrial projects appeared as a publication during the next five years.

The first Board of the Faculty in 1909 consisted of the Dean[3], the Vice-Chancellor (Sir Isambard Owen) and all eight Professors, together with Mr. J. C. Inglis (later Sir James), who was not only President of the Institution of Civil Engineers at that time, but also General Manager of the Great Western Railway. He died in 1911, but his place on the Board was taken successively by Sir William H. White FRS (1911–13) – an eminent naval architect, the Right Hon. Sir William Mather (1914–20) – a mechanical engineer and textile manufacturer, and the Hon. Sir Charles Parsons FRS (1920–28) – a polymath engineer who invented electrical machines and founded the firm which carried his name. The records

---

[3] This order in which board membership is listed, with the Dean taking precedence over the Vice-Chancellor, is that recorded in the Faculty Prospectus. It gives further indication of Wertheimer's view that he 'was Master in his own house'.

indicate that they were appointed with the remit '. . . to shape the contents and style of the university course so that it related to the real engineering world'. From subsequent developments of the Faculty it is clear that these four gentlemen approached their responsibilities with sensitivity and far-sightedness, persisting in an emphasis on principles, rather than the immediate needs of current practice in any of the various branches of engineering which they represented.

In 1927 the Board, with Andrew Robertson as Dean, invited four non-professorial, but senior, members of the academic staff to join it, and when Sir Charles Parsons retired in the following year, he was not replaced by any external member. The composition of the board continued in this mode until 1971, but with the Dean now formally taking over the chair from the Vice-Chancellor, and with an increasing number of appointed non-professorial members as the Faculty grew in size. In that year it recommended to Senate '. . . that all permanent members of the Faculty academic staff be permanent members of the Board', but it was not prepared to take the next step of accepting student members, commenting

> . . . no good purpose would be served by electing student
> members to the Board, it already has a Staff-Student
> Consultative Committee (SSCC), an Engineering Society, and
> an efficient personal tutorial system.

However, it could not hold this position against the general acceptance of students on the boards of other faculties, and in June 1975 it approved the proposal 'That 2 students of the Faculty be members of the Board, one to be the Faculty student representative on Senate; the other to be nominated by the SSCC.' But it was not until November 1984 that it agreed to have two elected representatives from postgraduates and research assistants.

A very noticeable feature of the Board's Report to Senate – from the beginning to the present day – has been their brevity compared to those of other faculties; they eschewed details of discussions, giving simply the conclusions of each item of business. This terseness was such, that in 1974 a request was made formally in Senate, that 'The Engineering Faculty be required to present a fuller account of its doings,' but the matter was not pressed – perhaps the University administration wished that other faculties would follow this example!

4.4 The First World War, 1914–18

The First World War which began in August 1914, had been anticipated earlier, and during the 1913 session, with Viscount Haldane as its second Chancellor, the University established a 'Military Curriculum' described as 'a course to persue for men between 16–21 in order to obtain a Commission in the Army or the Indian Army.' Cadets, as such students were called, joined the University Officers Training Corps (OTC) and were required to be 'British subjects of pure European descent'. In addition to their chosen degree courses in engineering, studies in military history, tactics and strategy, field engineering and map reading were mandatory; and as well as OTC camps, six weeks' attachment to a Regular Army Unit was required. Because it was still the norm at this time to buy a Commission, the inducement offered to young men was that this route to officer rank would '. . . for those able to live at home costs would be far less than obtaining a Commission through Sandhurst or Woolwich.'

The actual beginning of the war had important consequences for the development of the Faculty. Although 76 students had registered, 21 of them withdrew to enlist in the Armed Forces having passed the first-year examinations which gave them exemption from the Army entrance examination, and four teaching staff had done likewise. The students remaining were joined by young Belgian refugees who were not fit for military service, whilst the remaining academic staff also taught in local secondary schools as replacements for absent teachers. As a further aid to the war effort, a course was started in July 1915 '. . . for the purpose of training educated women to gauge munitions'; presumably this meant what we would now refer to as quality control.

As the war progressed the Faculty Reports include lengthening lists of former students of both the MVTC and the Faculty who had been either killed or wounded, and shorter lists of those who had been awarded medals for gallantry in action. The number of active students fell to 35 (with 38 on war service) and only the Professors were left to do the teaching. By this time the munitions classes had grown to 304 men and 233 women, a burden which became unsupportable, causing the classes to be stopped. Instead, the engineering workshops were made over to two-month training courses for men of the Royal Flying Corps. Research effectively ceased, with the disabled[4] Prof. David Robertson as the only author of publications of an engineering nature although Prof. Wertheimer continued to publish on educational topics, such as an article in the science magazine *Nature*, under the title 'Science and Modern Languages in Civil Service Examinations'.

Towards the end of the war the government required universities to relax entry requirements for ex-servicemen, and the Faculty played its part in offering rehabilitation opportunities for both officers and other ranks. For the former, the teaching was described as 'Officers' University and Technical Classes' but no details were given of their content. For disabled soldiers and sailors an 'Electrical Handyman' class was made available by Prof. Robertson, as was a 'School of Wireless Telegraphy' for sailors who intended to take employment in the Merchant Navy. Of note also is that during the 1917 session a series of lectures on aeronautics was attended by 2700 (*sic*) people; presumably this rather large number was the aggregate of numbers attending the complete course of lectures, unless the typesetter had included one too many noughts!

4.5 The Immediate Post-War Period

There were three significant events during the 1918 session, the first after the war. What turned out subsequently to be the most important of these for the future of the Faculty was that, at a salary of £800 p.a., Major Andrew Robertson DSc replaced Prof. Munro who had joined the MVTC 48 years earlier. The military title of this new Professor arose from the fact that at the outbreak of war he had been required to leave his post at Manchester University and to join a research team at the Royal Aircraft Establishment, Farnborough, where all senior researchers were given military rank so as to allow them to interact with the Armed Forces. Robertson's influence on the development of the Faculty from these early days until his retirement in 1949 will be described fully later in this history; it can hardly be overstated. The second of the significant events was that the number of registered students had risen to 181, although 107 of these had remained with the Armed Forces during part of the session. Finally, in a section described as 'Business Training' the Annual Report for 1918–19 notes that

---

[4]   This was not a war injury; he had been confined to a wheelchair for most of his life.

In the United States and Germany engineering undertakings
are managed much more generally by engineers than is the
case in this country; this is largely due to the fact that
American and German engineers receive some training in
commercial subjects.

It goes on to report

that the Faculty Board had obtained the approval of Senate to
include as part of the curriculum for the BSc in Engineering
degree, a course which includes lectures on bookkeeping and
Accountancy, Works Administration and Commercial Law.

It is unlikely that anyone present at the time would have doubted that
some basic knowledge of these subjects was essential for the successful
professional engineer, but there must have been concern by some of
the academic staff that the introduction of these subjects was not consonant with the previously stated aims of the Faculty in providing the
scientific basis required by the profession, leaving the practical training
to others. Perhaps Wertheimer was trying to introduce some of his own
MVTC ideas into the syllabus before Andrew Robertson could make his
mark, but he failed to do so because there is no reference to these
courses being introduced during this early period of the Faculty's
history. Instead, as will shortly be described, a sandwich scheme of
training was started.

In this session the Faculty began recruiting new staff, but whether by
inclination or government direction consequent on the acceptance of
exchequer grants, and its wish to find employment for ex-servicemen,
the advertisement for a lecturer in electrical engineering contained the
following passage: 'preference will be given (a) to teaching and works
experience, (b) service with the colours, (c) been disabled whilst so
serving provided his health and energy are not impaired.'

The 1919 session saw a step-change in the number of students to 301,
although 49 of these were taking a draughtsman's course for disabled
soldiers. The annual course fee had risen to 30 guineas, but a large number of scholarships were available to cover this fee, amongst which were
ten offered by the Society of Merchant Venturers for '. . . the sons of
officers killed in the war, whose mothers or guardians are in needy circumstances.' The restriction here to sons of officers speaks volumes

about the social stratification existing in the UK at this period in our history. This restriction to the officer class continued until the 1926–7 session, after which the awards were not made, but a year earlier a single new scholarship was made available to 'a candidate whose father was a student at the College and had lost his life whilst serving with HM Forces during the Great War.' By the 1932–3 session the Society were offering six scholarships, without restriction, to cover tuition fees.

The expansion in student numbers necessitated new laboratories being built for Mechanical and Electrical Engineering in nearby Orchard Street, effectively doubling the available space, and grants of £7000 and £3000 were obtained for equipment from University and government. No doubt with an eye to the future, a lecturer in aeronautics and physics was appointed.

During the First World War, which, like most wars, caused a break-up of the existing social fabric, various politicians, most particularly the Coalition Prime Minister, David Lloyd George, promised the country and the returning members of the armed forces, that a 'Land Fit For Heroes' would be created after the war ended, but this nirvana was not realised, and by 1921 unemployment and the generally low standard of living caused a national crisis and the creation of a military-style body described as the National Defence Force. Because the principal purpose of this body was to defend the coal mines, without which industry would come to a standstill, the Faculty's important educational activities in mining were targeted as a source of recruitment for this defence force. A letter from the Board of Education to the Vice-Chancellor set out the government's attitude: 'Every facility should be offered to students to join the Defence Force or to serve in any other organisation established for the protection of national interest in the present emergency.' A Government-inspired suggestion was that the summer term that year should be cancelled, but Oxford University refused to do so, and, through an advertisement in *The Times*, succeeded in persuading others to follow.

A spate of letters between the Vice-Chancellor and Prof. Wertheimer followed, chiefly on the 90-day period of duty required of the volunteers, questioning whether the University would (or could according to Regulations) excuse students from classes for this period, and whether the various scholarship awarding bodies would respond positively. The written correspondence shows that Wertheimer was his usual acerbic self in these exchanges, but Owen contributed significantly to the difficulties of decision-making because, in his final year before retirement[5], he was not often in the University to make the required decisions. For example, a letter from the Manager of the Lydney and Crump Meadow Colliery in Gloucestershire, asked that his son, who was a student in the Faculty, be allowed to stay at home to keep the steam pumps working, without which his mines would become flooded – his 640 miners being on strike! In the Vice-Chancellor's absence, Wertheimer took it upon himself to write a letter giving the son permission to be absent from his studies, and it must be assumed that he similarly excused all the other students who joined the Defence Force, particularly as the majority were ex-servicemen anyway. If University regulations were to be breached, he left it to the Vice-Chancellor to sort out!

Returning now to 1918, we find Wertheimer once more doing his utmost to promote the interests of 'his' Faculty, this time in the matter of finance. The occasion was a letter which Owen had received from Sir Oliver Lodge, then Chairman of the Committee of Vice-Chancellors and Principals (CVCP), concerning a proposed meeting called by the Board of Education at the invitation of the Chancellor of the Exchequer. Its agenda would not be out-of-place today, because four of its five points for discussion were – increase in staff salaries, additional staff, further development of research, and reduction of student fees; the fifth topic was related to the resolution of post-war difficulties. At this time, Board of Education grants were made to Faculties of Arts and Science, with Engineering considered as part of the latter. Wertheimer pointed out to Owen that Bristol and Manchester were special cases, in that their Engineering Faculties were financially autonomous, and that he, Owen, should make a statement at the meeting on 'grants available in aid of technological and professional work in universities'. It is not known whether anything came of this, but what we do know is that the meeting resulted in the University Grants Committee (UGC) being created in the following year (1919).

One final illustration of the correspondence between these two men concerned Bristol's contribution to a history of the support which UK universities had given to the war effort. As published, it contained no reference to the Faculty, causing Wertheimer to itemise what he thought should have been included, the most important of which was

---

[5] Dr Thomas Loveday replaced him as Vice-Chancellor in 1922.

the information that Prof. Morgan had spent the war helping Mr Mills[6] to perfect his eponymous bomb! Owen could only give the limp reply that he was not responsible for the final contents of the book.

4.6 Educational Links with Industry

It was currency inflation which had caused the course fee to rise to £42 per annum in the 1921 session, with additional fees for courses such as surveying and draughtsmanship. On the positive side, participation by engineering students in the Guild of Undergraduates (later the Students Union) was noted with approval, particularly that three of the last four Presidents had been from the Faculty, and that it had won the Inter Faculty Sports Cup, an achievement which was to be repeated in a number of succeeding years. An athletic ground of 26 acres (later extended to 33) had recently been provided by a gift from Sir George Wills, then Chairman of the University Council, and in 1920 the same benefactor presented the Victoria Rooms as a home for the Guild of Undergraduates. In the matter of examination results, perhaps for comparison with today's statistics, of 51 candidates 31 were successful; there were three first-class honours and seven second class, whilst eight of those who failed were awarded the University Certificate in Engineering. It does not say what happened to the remaining 12!

The 1921 Report is also noteworthy in that it records the satisfaction of engineering firms with the 'Bristol Sandwich Scheme of Training'. This had been introduced the session before as an option lasting a full five years with intercalated periods of education in the University and training in industry. The first 10 months were to be spent in the University and the next 14 in industry, and to deter any student who might think this would be an easier option, they were informed that

> Students of promise and excellent character can avail
> themselves of this arrangement, but if any student does not
> give satisfaction during the first 14 months (the first industry
> period), he will not be able to return to the works during the
> subsequent periods . . . and he will of course be liable to
> dismissal for idleness or misconduct.

---

[6] The Mills bomb was a hand-grenade, millions of which were used in the war.

A total of 23 firms participated in this scheme for the first five-year period, paying the student between 5 and 15 shillings per week when working there, and reduced premiums were offered to students who subsequently registered with the firms as apprentices or pupils. The success of this sandwich training scheme is commented upon in several succeeding reports, but no details are given of how the arrangement was managed, and what roles, if any, was played by the Faculty academic staff during the periods in industry.

The seriousness with which foreign language teaching was taken was emphasised at this time by the following statement in the course programme: 'Candidates for the final examination shall give evidence of their ability to express in English the meaning of passages from scientific publications in French and German.' How this ability was actually to be measured, and at what level, was not stated.

4.7 Professor Andrew Robertson as Dean, 1924–49

The death of Julius Wertheimer in August 1924 marked the end of the first period of the Faculty's history, and the appointment of Andrew Robertson as his successor as Dean began the transition of the Faculty from one in which excellence in teaching was the primary concern, coupled with technical assistance to mainly local industries, to a Faculty in which research took its place as of equal importance to teaching. He was also concerned to make a clean break with the Wertheimer period in the Faculty's interaction with the University. When the Registrar (Geoffrey Francis) showed him the draft of a history he was writing of that time, Robertson insisted that he remove his references to 'acrimony', and that he should let 'bygones be bygones'.

As we shall see, this emphasis on research, introduced by Robertson, resulted later in attracting a succession of academic research engineers who led the Faculty to national eminence in all its activities, and to international acclaim in several. It also resulted in a change to the BSc(Eng) regulations; in the future there were to be two classes of honours, first and second with two divisions in second class. This was clearly a mechanism for distinguishing the very good from the rest, but the reason for a second change is not at all obvious; it is recorded as follows:

> Candidates who have taken a degree with honours in one
> branch of engineering may become candidates in another

branch after attending for one additional year at least such courses in the university as the Engineering Board may prescribe.

Was there a serious demand from students, or from industry, for a broader engineering education?

During Wertheimer's long period as Dean (1910–24), his annual reports were printed in octavo, booklet format, and were clearly intended as public documents. But Robertson's were typed on cheaper foolscap paper, for internal record only, and by 1925 there was a decided difference in content. In a prefatory note he describes the history of the Faculty up to that time, and then remarks 'At the present time this College is during the day devoted entirely to the work of the Faculty of Engineering, except for one or two classes in commercial subjects.' [7]

This is clearly intended to mark a distinct break between the Faculty and the MVTC. Another break was the appointment of four lecturers to the Faculty Board, which since 1909 had consisted of Professors only.

Although the usual data about students was recorded – their origins, courses, examinations and employment – a large part of the report of 14 pages was taken up by detailed descriptions of research in progress in the different departments and work for industry which now had a national, rather than local, dimension. Of special note is the fact that Robertson himself had been awarded the Telford Gold Medal of the ICE for his paper on 'The Strength of Struts', a research area which was to occupy much of his life, and which was to bring him international recognition as the creator of the 'Perry-Robertson' formula which is still used today.

Prof. Ferrier retired in 1928. He came to Bristol in 1901 as Head of Engineering at UCB, becoming the first Professor of Civil Engineering when the Faculty was created in 1909. Although the annual reports record him as contributing greatly to testing work for industry, and some research on the properties of concrete, his valedictory entry in the 1927 Report appears to sum up his contribution by ' he has been a most successful teacher'. He had certainly found it difficult being second-in-command to Wertheimer! Fig. 4.1 is a photograph of members of the University taken sometime between 1919, when Andrew Robertson was

---

[7]   A lecturer in 'Business Knowledge' was appointed in 1927.

Fig. 4.1 A photograph (*c.*1921) of the whole University. The Vice-Chancellor, Sir Isambard Owen is prominent in the centre foreground, and immediately behind his left shoulder are Profs. Arthur M. Tindall (Physics) and Wertheimer; at their backs are Ferrier and Andrew Robertson.

appointed, and 1922 when Sir Isambard Owen retired: Owen occupies the front centre ground, with Wertheimer (with moustache) to his left and slightly behind. Directly behind Wertheimer is Robertson, and to his immediate right, Ferrier. The man between Owen and Wertheimer – but slightly behind – is almost certainly Prof. A. M. Tyndall, a leading member of the physics department and acting Vice-Chancellor before Sir Philip Morris was appointed in 1946.

At first sight, a surprising fact is that for the following year the number of students in the Faculty was 83, only one more than 20 years earlier, but the likely explanation has already been alluded to, the deliberate increase in research activities, requiring academic staff time and experimental facilities. For example, in this session (1928) Robertson had persuaded A. J. S. Pippard[8] (Fig. 4.2) to transfer his Civil Engineering Chair from Cardiff to Bristol. It will be remembered that

[8] In 1920 he had obtained a Bristol DSc, and as a result of wartime research at the Air Department of the Admiralty he was awarded the MBE and published a book on 'Aircraft Structures'.

Fig. 4.2 Alfred John Sutton-Pippard (on the left) and John Fleetwood Baker (on the right), the two inter-war Professors who established the tradition of high-quality research in the Faculty.

Pippard was the best student to graduate in the Faculty in the 1910 session and, building on that, his research reputation was already high, specifically in the experimental and theoretical aspects of aircraft structures. Later, when he moved to Imperial College, it was further enhanced through basic research on strain-energy methods[9] in structural analysis, leading to his election to the Royal Society in 1954. To facilitate Robertson's research plans for the Faculty, he had himself designed a 50-ton testing machine for compression and bending tests on structural members, to be funded by a UGC grant and manufactured by the Denison Company. With modifications, it was used 20 years later by Prof. Pugsley's research group working on aluminium structures, and was still in use when the author joined the faculty in 1956.

By 1930 the international status of the Faculty had begun to be established, evidenced by the fact that it had been possible to organise practical training of students not only in the Bristol Aeroplane Co. and Metropolitan-Vickers in the UK, but also with the Aluminium Co. of

[9] The principles of strain-energy methods were first produced by an Italian, Castigliano, in 1879, and Pippard records that he was introduced to them by Frank Broadbent, one of the civil engineering lecturers in the Faculty when Pippard was a student.

Toronto, the Indian Public Works Department and the Allgemeine Elektricitals Gesellschaft of Berlin. But the annual report is now less concerned with students and the current employment of former students, than with the activities of individual departments and their academic staff. We find, for example, that Profs. A. Robertson and Pippard were members of various committees of the Department of Scientific and Industrial Research (DSIR) where Stradling (see 4.2) was Director of Research; that Prof. D. Robertson was Assessor for National Certificates in Electrical Engineering, and that Prof. Morgan was currently President of the Institution of Automobile Engineers. A gentleman by the name of Mr. P. E. Gane offered the Faculty funds for an enigmatically titled 'Chair in Fire Appliances', but nothing came of it. By far the most observable new feature of the annual report is the space given over to research activities, and as an illustration of the detail presented, interesting because it records the state of computation 80 years ago, we find the following: 'After the first paper [by Pippard] had been presented to the Aeronautical Research Committee a grant was made for the purpose of employing a computer for a period of 6 months.' Perhaps the recording typist made a spelling mistake here, because the computer was not a machine, but a lady named Miss White! Although he was a Professor in the Civil Engineering Department, Pippard was a member of the Aeronautical Research Committee and its Airships Sub-Committee, participating, as many eminent structural engineers did at that time, in research relevant to airship and aircraft structures.

It would seem that at this time, having first survived, and then begun to prosper during its first 20 years, the University was beginning to plan seriously for its future because the question of 'Urgent Needs' became a required consideration by each faculty. In 1930 Engineering had two requests, the first being for additional staff so that a four-year (from matriculation) honours course could be instituted; the second was for development of a course in building science, also requiring more staff and equipment. Additional staff were certainly recruited in 1932 but they were used ' . . . to bring the Honours School more into line with the more progressive universities and to enable the teaching of certain stages of pass and honours students to be separated.' This was certainly achieved because students entered the Faculty without preference for a particular department and continued so until the beginning of their third year. As a result of first-year examinations they were divided into 'degree' and 'diploma' groups which took different examinations at the end of the second year. The recorded numbers show that the 'degree' students were then divided further into 'honours' and 'ordinary' for the third-year courses and examinations, as a result of which the best students were allowed to proceed to a fourth year examination for an honours degree, whereas others were given an ordinary degree as a result of the third-year examinations. Those relegated to the second stream as a result of the second-year examinations proceeded to take the diploma examinations at the end of their third year.

In case intending students should be uncertain as to the principles upon which the course content was to be based, from this period and for many years subsequently, the Faculty Prospectus contained the following reiteration of those principles first set down when the Faculty was created in 1909:

> The object of the course in engineering is to provide instruction in those principles of science, the knowledge of which is necessary to an engineer in his work. They are designed to give a broad rather than a specialised training and so may be found suitable for some who require a general scientific training but do not intend to take up engineering as a profession. The courses are not intended to supersede the usual routine of engineering training in workshops or upon civil engineering construction. An apprenticeship, pupillage, or the equivalent, after a period of university study, is essential to everyone who proposes to enter the profession of engineering.

Three of the four new staff appointed in 1932, W. H. Dearden (metallurgy), F. de la C. Chard (electrical) and S. T. Newing (mathematics) remained in the Faculty for many years until long after the move to Queen's Building in 1955. Although, as noted above, the second Faculty 'urgent need' after recruitment of additional staff had been the introduction of building science into the curriculum, no progress was made in this direction.

In 1932 Pippard was attracted to a Chair at Imperial College, London, but another bright star in the structural engineering firmament, J. F. Baker (Fig. 4.2), replaced him in 1933. Baker, then aged 32, had earlier been an assistant to Pippard at Cardiff, and was then working at the Building Research Station; he was also Chairman of the Steel Structures Research Committee of the DSIR, Andrew Robertson and

Pippard being members. The work of this Committee was recognised by these three as being of such importance that in the three sessions 1937–40 the Faculty offered a postgraduate course in structural analysis and design, '. . . prominence being given to the work of the Steel Structures Research Committee and the developments resulting from it.' As the originator and developer of the plastic design method for steel structures, Baker was to reach an eminence in this field possibly greater than Pippard's. As early as 1933 he had been awarded the Telford Gold medal of the ICE, becoming a Fellow of the Royal Society in 1956, Professor of Engineering at Cambridge (1943-68) and Lord Baker of Windrush in 1977. As a peer, he was for many years active in injecting an engineering dimension into the business of the House of Lords, most notably on educational matters.

During the remainder of the 1930s, student numbers were maintained at around the 80 mark, with continuous improvement of quality as measured by the percentage having matriculation. But with Baker, W. M. Shepherd and F. de la C. Chard leading, the volume and quality of research continued to grow. Sadly, with war having been narrowly avoided in 1938, but still a hazard to be prepared for, Baker was seconded to the Civil Defence Research Committee, initially part-time, but later full-time as the war started, becoming Scientific Adviser to the Ministry of Home Security.

4.8 The Second World War Period, 1939–46

There were many changes to the Faculty immediately after the start of the Second World War on 3 September 1939. It agreed to accept the engineering faculty of King's College London, and since this included both staff and students, the former were in some part able to replace the Bristol staff who had joined the Armed Forces, notably Lt.Col. Elgood and Capt. Chard, the former to the Royal Engineers and the latter becoming Chief Signals Officer for the South of England. The Royal Naval College at Greenwich was also evacuated to Bristol and the Faculty provided it with laboratory facilities, their teaching needs being dealt with elsewhere. The laboratories were also used by the Aeronautical Inspection Directorate, and the workshops as a training centre for the Ministry of Labour.

The number of registered students was 90, but 12 of these had interrupted their studies to join the Armed Forces. Surprisingly perhaps, the nationalities of students was even wider than usual, Burma, Malay States, Moravia and Turkey being amongst a long list of countries of origin. The general welfare of these students at this time was the responsibility of the Colonial Students Welfare Committee – so named until 1951 when it became the Overseas Students Committee. Not surprising, was the Fellowship of the Royal Society accorded to Prof. Andrew Robertson, and Prof. P. M. Dirac (a former electrical engineering student) was awarded its Royal Medal for the leading part he had taken in the development of quantum mechanics. As previously noted, his father, Charles Dirac, had taught French to engineering students in the early years of the Faculty.

In 1941 Prof. D. Robertson died; he had been in Bristol since 1902 and was the first Professor of Electrical Engineering. This year also saw the post of senior lecturer created in the University, its first recipient in the Faculty being F. R. B. Watson of Mechanical Engineering. Student numbers had dropped to 62 and research publications numbered only two. This was not surprising because five of the academic staff, Morrison, Shepherd, Gibbs, Newing and Dearden, spent their vacations assisting on various technical problems at the Bristol Aeroplane Co. Shepherd was also an ambulance driver during the period of heavy bombing on Bristol, which in 1940 destroyed the tower of the Great Hall of the Wills Memorial Building. In his later years he needed little encouragement to re-tell the story of driving an ambulance down Park Street effectively on three wheels, the fourth having sheared its connecting bolts in bumping over a bomb crater! In 1946 he was the third[10] member of the Faculty staff to be appointed to the Readership title.

The very serious war situation in 1942 resulted in the decision to discontinue publication of the Faculty Prospectus in its usual form; instead, a single sheet was issued describing several enforced changes to its activities. Every physically fit male student over 18 was required to become a member of either the University Senior Training Corps or the University Air Squadron, and together with the academic staff, to take part in the Air Raid Precautions (ARP) fire-watching and fire-fighting services of the University, and '. . . to give such unpaid services as may be required'. The liability for this service extended over the vacations,

---

[10] Because geology was part of the Faculty until 1950, its first Reader (in palaeontology) was Stanley Smith, appointed sometime during 1942–45. The second was J. L. M. Morrison (Mech.Eng.) in 1945.

but arrangements were made by which students whose homes were at a distance from the university could do their share of fire-watching during term-time. Exceptions to these rules were given to conscientious objectors, subject to approval by the tribunal set up nationally to deal with such people, and also to students who were members of the Home Guard[11] – a voluntary body of ordinary citizens who were either over military age or in reserved occupations, who received training in their spare time in basic fighting techniques in order to resist an invasion of the UK by German forces. It is interesting to note that exemption from ARP duties could be granted by the Committee of Deans '. . . on the grounds that a student's course of study is so exacting as to make compliance with the regulations undesirable.' Unfortunately, no examples are given as to which courses might qualify, but presumably medicine was one such.

In order to accommodate the King's College faculty of engineering it was necessary to discontinue the final honours year, and, as a war-time measure, to award honours on the results of the third year honours examination. Also, to meet government demand for mechanical and electrical engineers, the civil engineering course was effectively discontinued, as was the specialised radio electrical engineering course, with some radio introduced into the general electrical engineering course.[12]

For the next two sessions the Faculty continued to survive, notable events being the election of Prof. Andrew Robertson to the council of the Royal Society, and the direction of graduating students into either war industry or the armed forces. Unlike the First World War, it was not thought necessary to record long lists of casualties, but the names of students who had been decorated for valour were recorded.

4.9 Preparation for Peace; New Leaders for the Faculty

In the immediate post-war period all members of the Faculty recognised the exceptional effort necessary to deal with the large number of returning ex-servicemen whose studies had been interrupted by the war – in some cases voluntarily – and who had been promised the opportunity

---

[11] The television series *Dad's Army* gives a humorous slant to the activities of this body, but the present author can testify – as a Home Guard 'runner'– to the seriousness with which it performed its duties.

[12] The Faculty missed an opportunity here because the Admiralty Signals Establishment moved into the Physics Department to help Watson Watt with the development of RADAR.

for their renewal when the war ended. But whereas the student/staff ratio pre-war had been around five, it was now 11, and the available academic staff also had the additional task of re-building the Faculty's activities as quickly as possible. As after the 1914–18 war, amongst these ex-servicemen there was a high failure rate in the intermediate and first-year examinations, but this was not wholly due to teaching staff shortages. These students were of varying quality in the academic sense, and many would probably not have been offered places in normal circumstances. It was also the case that some had experienced traumatic war-time situations – perhaps being seriously wounded or as a prisoner-of-war – which made dedication to study extremely difficult.[13]

The bright future for the Faculty appeared to dim in 1943 when Prof. Baker resigned from his Civil Engineering Chair to accept the Chair of Mechanical Sciences at Cambridge. Fortunately the dimming was more apparent than real, because in the same year G. H. Rawcliffe, then at the University of Aberdeen, was appointed to the Chair of Electrical Engineering, to be followed shortly after by A. G. Pugsley to Baker's Chair of Civil Engineering. Both were to be instrumental in establishing their Departments at international level in research, partly through their own personal contributions, but also by their ability to choose highly effective colleagues. Unlike Pugsley, Rawcliffe had hardly begun to earn his reputation when he was appointed at Bristol, but his work on alternating current machinery resulted in PAM – pole amplitude modulation, a topic to which he was to devote the greater part of his academic life and which gained him Fellowship of the Royal Society in 1972, just five years before he died[14]. He was a prominent member of the Institution of Electrical Engineers, becoming its Vice-President in 1972–75, but ill-health, regularly induced by bouts of asthma, caused him to decline the Presidency.

Pugsley, on the other hand, had already established an enviable reputation at the Royal Aircraft Establishment, Farnborough both before

---

[13] The author started his university studies at Imperial College in 1947 and learned to appreciate the difficulties faced by many ex-servicemen in returning to university life. In return for an education in 'life-skills', he was able to reciprocate by helping them remember the basic sciences which they had forgotten.

[14] His Royal Society Biographical Memoir states 'The breakthrough came in Rawcliffe's work with R. F. Burbidge on 2:1 pole-changing windings. Rawcliffe had given Burbidge (a Res. Asst.) a paper by a Frenchman (Dahlander) to read, and it was Burbidge who saw its implications, but it was Rawcliffe's overall grasp of the problem which enabled him to make the practical application'.

and during the war. As much as any other individual researcher he contributed to its successful outcome, particularly the Battle of Britain in 1941, by his work on the flexural-torsional coupling of beams, which persuaded R. J. Mitchell, the designer of the 'Spitfire', to move its main wing-spar backwards so as to increase the torsional stiffness. Sir Sidney Camm was persuaded to do the same for the 'Hurricane', and it was these two fighter aircraft which defeated the German Luftwaffe in the 1940 Battle of Britain. In 1940 he became head of the Airworthiness Department at Farnborough, a post which turned his interests towards the concept of structural risk and safety, areas in which he was to make himself, and Bristol, internationally famous. He had been enticed by several universities at home and abroad to accept chairs, but on the advice of such eminent colleagues as Southwell[15], Pippard and Baker, he chose Bristol. His contribution to the Civil Engineering Department, the Faculty and the University – he served as Vice-Chancellor in 1963 during the illness of Sir Philip Morris – were such as to earn him an Honorary Fellowship of the University in 1986, the highest honour it can bestow. He was elected to the Royal Society in 1952, made Knight Bachelor in 1956, and was President of the Institution of Structural Engineers in 1957–8.

If the reader of this history should wish to know more about the academic, engineering and personal characteristics of Pugsley, I can do no better than to suggest reference to the Royal Society Biographical Memoir[16] written by one of his earliest research student, Henry (now Lord) Chilver. His students recall being grateful for the careful preparation of his lectures, although they were delivered in such a soft voice that there was always a rush for the front seats! On personal matters, he had a formality in his character which meant that outside his immediate family, no one – not even his fellow Professors – referred to him personally by anything other than his surname, and they were amused by his retention of Civil Service protocol, which not only required him to have a carpet on his office floor, but a mat outside his door – the only one in Queen's Building! However, his catholic range of interest and achievements, coupled with a good memory, resulted in animated and

[15] Sir Richard V. Southwell, Professor of Engineering at both Oxford and Cambridge; also Rector of Imperial College.

[16] The same would be true for Andrew Robertson, Collar and Rawcliffe. For Morrison the IMechE obituary notice could be consulted, and for Shepherd the obituary notice in *The Times* newspaper.

wide-ranging conversation – particularly in mixed company! It is to this last feature of his character that the University owes the greater part of its collection of the artefacts of I. K. Brunel. From the early days in Bristol, Pugsley's habit every Wednesday afternoon, was to take his wife Kathleen for shopping and tea in Bath. My recollection of a much later conversation with him, is that one of the several ladies whom they sometimes met for tea was Lady Celia Nobel, a grand-daughter of IKB, who made it known to Pugsley that in her possession were several boxes of IKB's papers, drawings and instruments. What is certainly true – the University Library (Special Collections) being in possession of relevant letters – is that in 1950 Pugsley, accompanied by Prof. W. F. Whittard (Geology), was invited by Lady Nobel to inspect these artefacts, a result of which was that they were subsequently gifted to the University as the initial part of its Brunel Collection.

The 1943 session also saw the appointment of Andrew Robertson to membership of the University Grants Committee (UGC), which gave him greater knowledge than most about possibilities for post-war developments. In Bristol he had experienced the overcrowding of the Unity Street buildings for 20 years and its restrictions on the development of the Faculty. His wide experience also persuaded him of the need for a national expansion in engineering education after the war. Initially, the UGC was providing funds to repair war-damaged buildings and the restoration of buildings which had been, or would be, returned to their owners after requisition for military purposes. When it later turned to the provision of new buildings it had in mind that the eventual ending of the war would require the government to honour its pledge to the young men and women whose education it had postponed by conscription into the armed services. Within the University, Robertson had already used his position as a senior Professor to ensure that priorities for new buildings were halls of residence, an engineering building, and a medical school – in that order. He prompted his Faculty colleagues to plan for the future in terms of accommodation and equipment for a new building, but it was not until February 1947 that the UGC formally asked the University to state its requirements, as a result of which the architect Sir George Oatley was asked to draw up plans and to survey the site (Fig. 5.1) of what is now Queen's Building. However, the UGC response, received at the end of March 1947 was disappointing; it allocated only £343,000 for a Bristol building programme over the next five years, which was hardly sufficient for an engineering building alone. Although planning continued, confidence could not have been high because as late as May 1950 the General Purposes Committee minutes record, referring to the proposed site of the new engineering building '.. that at some future time the possibility of using the ground for staff tennis be considered'.

Returning to 1946–7, it is noted that student numbers had risen to 179, two-thirds of whom were receiving scholarships from either the state or local education authorities. Although graduating students were being directed either into strategic industry or the armed forces – a two-year period of National Service being required – a new section appears in the Dean's report describing the research being carried out by postgraduate students, much of it being on the subject of the behaviour of light alloy structures being used in aircraft, a topic for which exemption could be obtained from National Service. One of the few positive effects of the war, was that at its end a considerable amount of unwanted materials testing and laboratory equipment became available for civilian purposes at almost no cost, and the Faculty took full advantage of this opportunity.

4.10 The Sir George White Chair in Aeronautical Engineering

The importance of Aeronautics as a new branch of engineering had emerged during the war, as new types of aircraft required structural engineering, and knowledge of materials, of a kind beyond that experienced by civil and mechanical engineers. It has already been noted that the senior academic staff of the Faculty had worked at the Bristol Aeroplane Co. (BAC) during their war-time vacations, and this, coupled with the fact that a 1924 Faculty graduate A. E. (later Sir Archibald) Russell was Chief Designer of Aircraft at BAC, would have contributed to the decision by that company to gift £6000 p.a. for ten years to the Faculty for the creation of the Sir George White Chair in Aeronautical Engineering. Its first incumbent was Mr. A. R. Collar, who will always be remembered for his early work at the National Physical Laboratory on the application of matrices to the solution of differential equations, particularly in dynamics, and the book[17], published in 1938 (and republished nine times) with R. A. Frazer and W. J. Duncan which

---

[17] Elementary Matrices and some Application to Dynamics and Differential Equations (CUP).

resulted from it. This basic research on aeroelasticity, particularly flutter, caused A. G. Pugsley, then Head of the Structural and Mechanical Engineering Department at the Royal Aircraft Establishment, to have Collar transferred there at the beginning of the war. It was no surprise therefore that Collar followed Pugsley to Bristol to create the Aeronautical Engineering Department, which in its first year (1945–6) had five students, rising to 25 three years later. At this time the Society still held the purse strings for the Faculty, causing Collar to remind his colleagues quite often that his early salary cheques came from the Treasurer of the Society, and not from the university.

Starting in the cramped quarters of Unity Street, Collar was given £284 in 1946 to erect partitions in order to establish his laboratory! His Department, and his research with it, flourished after the move to Queen's Building and he was elected to the Royal Society in 1965. But Roderick Collar – one of nature's gentlemen – was a man of many talents; a fine cricketer and footballer in his youth, a very polished raconteur and composer of comic verse which he often sang at the Bristol Savages Club, and, like Pugsley, an unstinting contributor to Department, Faculty and University. In this last role he became Vice-Chancellor in 1968 on the sudden death of John Harris, and during the 'student revolution' of that year his personal characteristics enabled him to steer the University through a difficult period, during which the University's business was carried on from the Vice-Chancellor's room in Queen's Building. They also ensured that he was elected to Honorary Fellowship of the University, but, sadly, he died in the very week that it was due to be awarded to him. In his memory, 'The Roderick Collar Prize' was established which carried the conditions fully indicating the character of the man himself:

> . . . to be awarded to second-year undergraduate students in
> the Department of Aerospace Engineering who have
> combined academic excellence with other activities. The
> Award will . . . be based on the students' academic
> performance over the first two years of the course, and on the
> contribution (e.g. societies and/or sport) made to the
> University and/or to Society in general.

The appointment of Rawcliffe, Pugsley and Collar within the space of two years was part of Andrew Robertson's ambitions for the Faculty, but the fourth member of this quartet, who was to contribute much to the creation of Queen's Building and of the Mechanical Engineering Department, was already there. John L. M. Morrison had been appointed by Robertson in 1928 as an assistant lecturer in Mechanical Engineering, becoming the second Reader in the Faculty in 1945, and Professor in the following year. His principal academic interests were in machine design, which during the war led him to serve on a wide range of committees concerned with guns and weapons design, and later on the Aeronautical Research Council and the Agricultural Research Council. Like Robertson, he found time to contribute greatly to the Institution of Mechanical Engineers, becoming its President in 1970, and like his mentor, was a man of unfailing modesty and courtesy, and an encouraging leader of his Department. He was the epitome of an educated Scotsman, with a concealed but infectious sense of humour, reluctant to speak unless he was the master of his subject, and seriously interested in golf! His other non-academic interests were fast cars – normally a Jaguar – tuned by himself – and apiculture. His involvement with Robertson in the design of commercial knitting machines will be discussed in Chap. 10.5.1.

To the foregoing quartet of Professors (Fig. 4.3) who were responsible for leading the Faculty forward – as will be described in the next chapter – should be added the name W. M. Shepherd. By the 1948 session student numbers had risen to 191 and the Dean's report contains the following:

> The position of Mathematics in the Faculty of Engineering
> has been formalised by the creation of the Department of
> Theoretical Mechanics of which Dr. W. M. Shepherd has been
> appointed Head.

The new department was created and Shepherd did indeed become its Head, but as will be reported in the next chapter, he did not become a Professor until 1959, and so any description of him will be deferred until then. It must be recorded, however, that this confirmation of the place of mathematics in Faculty teaching owes much to Andrew Robertson's firm friendship with the University's Professor of Mathematics, H. R. Hasse. They had been young colleagues at Manchester, and Hasse had followed Robertson to Bristol in 1919.

To teach and generally educate this increased number of students, the Faculty academic staff numbered 17, made up of three, four, four, two

Fig. 4.3 The Founders of the modern Faculty in the order in which they joined: Profs. Robertson, Morrison, Shepherd, Rawcliffe, Pugsley and Collar.

in Civil, Mechanical, Electrical and Aeronautical, respectively, with an additional three in Theoretical Mechanics and one in Metallurgy. For comparison with today's figures, the student /staff ratio was therefore a little over 11. Accommodation was extremely crowded in the MVTC, even though neighbouring houses in Orchard Street had been rented, and every possible space converted to provide workshops and laboratories. Even so, those who experienced these conditions – as undergraduate or postgraduate students – still comment on the exciting and friendly and productive atmosphere which had been created, enhanced without doubt by research discussions taking place in Carwardine's coffee shop, and Departmental visits to the cinema to see the latest films!

The University had a *locum parentis* attitude to its students at this time, normally requiring those who did not live with parents or guardians to live in halls of residence, of which there were three for men – Wills Hall, Burwalls and Wraxall Court, and two for women – Clifton Hill House and Manor Hall. For these halls a single room cost £94 for board and residence for a session of 30 weeks, with any additional week charged at 38s 6d. For comparative purposes, the course fee for the three-year degree was £166 16s 6d. The University also took a practical interest in the health of students, requiring them in 1938 to have a medical examination in each of the first three years of their studies. It is also on record in the same year as saying 'The University intends to provide a Centre near the Union and main University buildings facilities for physical training and for *quick-time* (my italics) exercises.' In 1946 it went further by continuing the free medical examination and requiring a chest X-ray for safeguarding against tuberculosis. When the National Health Service Act became law in 1949 it became possible for students to register with the University Medical Officer at his surgery in Woodland House, which by that time also contained squash courts and physical training facilities.

The 1948 Dean's Report, as well as recording that the Faculty had organised the Colston Society's Research Symposium on 'Engineering Structures', contains a section on 'Constitutional Changes' which records the reason why this chapter of the Faculty's history ends with the 1948 session. It is worth recording in full.

> This report must conclude by formally reporting the end of
> the long connection between the Society of Merchant
> Venturers and the Faculty of Engineering. As from 1st

September 1949, the Society of Merchant Venturers will no
longer be responsible for the administration of the College
and the Faculty, and the title Merchant Venturers Technical
College will cease to exist. The administration of the Faculty
will be carried on as for the rest of the University, and will be
entirely separate from that of the College. The premises are
being sold to the Corporation of Bristol, certain parts thereof
being leased to the University of Bristol for the use of the
Faculty of Engineering. Construction of new independent
premises for the use of the Faculty on the main University site
is shortly to begin. The College will be known as the College
of Technology and that part of it which is leased to the
University will be known as the University Engineering
Laboratories.

As a permanent reminder of this ending of the formal and financial link
between the Society and the University, a memorial tablet was installed
in the Faculty's new building, construction of which is described in the
next chapter. Fig. 4.4 shows the Chancellor, Sir Winston Churchill,

Fig. 4.4 The Merchant Venturers Technical College memorial tablet in Queen's
Building being unveiled by Winston Churchill in 1950.

unveiling this tablet, and because the text is not clear from the photo-graph, it reads as follows:

> This tablet commemorates the establishment and provision of
> a Faculty of Engineering in the University from 1910 to 1950
> by the Society of Merchant Venturers: Robert Sinclair, Master,
> 1953.

A further reason for ending this chapter here is that Andrew Robertson retired at the end of this session. The Faculty lost a powerful voice in University affairs, but, like all good leaders, he had appointed succes-sors who were collectively more than able to fill his place. They were pressing hard for a new building, and the agreement reached with the Society about the sale of the Unity Street building made it almost certain that a new home would be found for the Faculty.

Chapter 5

Building the Foundation for Excellence, 1950–75

5.1 Introduction

The previous chapter of this history ended with the retirement of Prof. Andrew Robertson as the second Permanent Dean, and the transference of complete responsibility for the Faculty from the Society of Merchant Venturers to the University. It was noted that Robertson had been responsible for the appointment of five Heads of Department, all of whom were to achieve personal goals of a high order, leading to national and international recognition for the Faculty. The last of these, G. H. Rawcliffe, retired in 1975, making that year an appropriate date on which to end this chapter.

The years immediately after the Second World War, which ended in 1945, were, if anything, more difficult in many respects for the civilian population than the actual war years[1], but the essential contributions made by engineers and engineering in winning the war indicated that education in all its sub-disciplines would require expansion of the existing provision. At national level the University Grants Committee had Andrew Robertson as one of its members, and it is unlikely that he failed to use his position to influence engineering education both nationally and in his own University in particular.

---

[1] Rationing of many basic commodities – such as bread – did not end until 1950.

5.2 The New Engineering School – Queen's Building

The prospect of new and expanded accommodation for the Faculty had been a priority in the mind of Andrew Robertson since he replaced Julius Wertheimer as Dean in 1925, and he clearly must have expressed his intentions to such outstanding researchers as Rawcliffe, Pugsley and Collar in persuading them to come to Bristol as soon – for the last two of this trio – as their wartime duties were nearing conclusion. The first recorded activity in this direction appears in the minutes of the University Developments Committee for May 1947 where '. . . plans for a new Engineering Building, to include Geology and Mathematics[2] were displayed and explained.' The site for this building was on undeveloped land (Fig. 5.1) which had been used during the war as allotments for producing food.

Robertson was a member of this Developments Committee and his arguments must have been persuasive because the Registrar was asked to set in motion consultation with the City Authorities. The Faculty Board was also asked to consider the plans.

In July 1947 the University architect, Sir George Oatley[3] and his deputy Ralph Brentnall – who had recently been released from military service – were asked to submit detailed plans. They did so in December that year and were given permission to start on-site investigation – to dig 'trial holes'– was how this activity was described.

By May 1948 the UGC had agreed the plans, allowing Brentnall to be instructed to prepare working drawings so that work could begin in September 1948. The reinforced-concrete frame, designed by the London-based structural engineers Mouchel and Partners, was to be clad principally in light coloured brick from Cheshire, with Bath stone for decorative features. But in this post-war period, not only did the Ministry of Works control manpower engaged in building and other strategic industries concerned with reconstruction, but it also controlled the supply of necessary materials. Following a meeting between the UGC and the Vice-Chancellor, an allowance of 200 men was agreed for the Engineering building and all other University

---

[2]   This was the University Department of Mathematics which was in the Science Faculty but Geology was in Engineering until 1950. As recorded in the previous chapter the Faculty had its own lecturers to teach mathematics to its students.

[3]   Also the architect responsible during 1912–25 for what eventually became known as the Wills Memorial Building.

Fig. 5.1 Queen's Building. The site, viewed from the South in 1945.

building activities. Not all these men were competent tradesmen, a fact which was to cause considerable delay, as did re-design of the columns because of the shortage of steel reinforcement.

The original design contained a large amount of ashlar facing so as to complement visually the Royal Fort and the Wills Memorial Building, but the UGC considered this to be an unwarranted expense. Even though the University found a generous benefactor to cover the cost of this ashlar, the UGC was not impressed and fixed a price limit of £623,000 for the first phase – the west wing and the central block. The final cost was £715,000 and, as we can see today, a pleasing amount of ashlar was actually used. The contractors were the Bristol firm of William Cowlin and Sons Ltd, and it was entirely appropriate that the first sod was cut by Andrew Robertson in 1949 operating a mechanical digger, with the Chancellor, Winston Churchill, laying the foundation stone on 14 December 1951. Fig. 5.2 is a photograph of the event; In the central seated rows of the audience, the front row is occupied by civic dignitaries, with the Vice-Chancellor, Sir Philip Morris, at the far end. But seated in the second row are, reading from right to left, the Registrar (Herbert Butterfield), Andrew Robertson (hatless), an unknown person, and then the four Heads of Departments, Pugsley

Fig. 5.2 Queen's Building. The Chancellor, Winston Churchill, laying the foundation stone (shown below). In the second row, reading from right to left, are Herbert Butterfield (Registrar), Andrew Robertson (hatless), an unknown person, then Pugsley (with hat), Morrison, Collar and Shepherd (all hatless).

(with hat), Morrison, Collar and Shepherd (all hatless). Rawcliffe was actually Dean at the time, but does not appear in the photograph, probably due to one of his serious asthma attacks. He was the first Dean to be appointed on a three-year fixed-term basis, but this recurring disability caused him to resign after less than two years in office, and eventually caused his death at the relatively early age of 69.

With the new building started, the Engineering Professors Committee began to prepare for the future, and its first step was to inform Senate 'That they wish to consider in due course a proposal for the establishment of a Chair in Theoretical Mechanics and for the appointment to the Chair of Dr. W. M. Shepherd.' The immediate result of this was the creation by Senate in 1953 of a committee 'to consider the general arrangements in the University for the teaching of mathematics', which took its time very leisurely to decide that the Faculty could continue to have a separate Department of Theoretical Mechanics, and not until 1959 was Shepherd made a Professor and continued as its Head.

The Faculty began its move from the MVTC in 1954 and virtually completed it by June 1955, the cost of moving all its machinery and experimental facilities being borne by A. E. Farr Ltd, a family firm of civil and structural engineers from Westbury in Wiltshire, whose part-ners had all obtained their engineering degrees from the Faculty.

The estimated cost of the second stage of the new engineering school – the east wing and the library – was £490,000 and construction started in November 1953, but the continuing shortages of both men and materials meant that it was not completed until the end of 1957 at a cost of £625,000, but part of this increase was due to ten extra offices being added to the second-floor of the already completed west wing. Fig. 5.3 is an aerial view of the completed building taken from the Royal Fort direction. The two car parks on the north side are to be noted; they were soon after improved in appearance by the planting of trees, and almost 50 years later they formed a key part in the re-development of Queen's Building during the BLADE project (Chap. 10.3). In further-ing their objectives of research and scholarship, the Professors placed the library in the most prominent position at the centre of the build-ing, with a magnificent outlook over the city of Bristol. By its detailed design, they ensured that, unlike many university libraries, it was, and still is, a place which actually produces the right environment for study (Fig. 5.4).

Fig. 5.3 Queen's Building. An aerial view from the north-west, taken in 1958.

After a preliminary visit by the Duke of Edinburgh in 1957, the new building was officially opened on 5 December 1958 by Her Majesty Queen Elizabeth II, and to commemorate the occasion it is called, by Her Majesty's consent, Queen's Building. Fig. 5.5 shows the Queen leaving with the Vice-Chancellor, Sir Philip Morris. Between the Queen and the Duke is the Pro-Vice-Chancellor and Professor of Medicine, Bruce Perry.

Although Andrew Robertson had been the initiator and guiding hand in the creation of Queen's Building, he retired in 1950 so that the detailed configuration and allocation of spaces for the different departments, which, it will be remembered, included Geology and Mathematics, was a task principally for the Professors in those departments, assisted greatly by Dr. E. F. Gibbs, the Senior Lecturer in Civil Engineering, for whom the new post of Assistant Dean was created in 1950.

Some of those who occupied Queen's Building during the second half of the twentieth century criticised its internal design, particularly the closely-spaced columns, which in the laboratories prevented large-scale experiments. The present author, who knew Professors Pugsley

Fig 5.4 Queen's Building. Library interior.

and Collar well, believes that these two eminent men, having been involved with large teams of researchers in major government laboratories, had concluded that a university researcher had a different role – the production of new ideas, integrated with simple experiments – and that if ever large-scale experiments should be absolutely necessary, access to existing laboratories outside the University could be arranged. Of course, they could not have foreseen that before the end of the century, many of these large government laboratories which they had in mind, would have been closed down, and neither did they anticipate where their own research might lead. In illustration of this, one of Pugsley's interests was wind loading on civil engineering structures, and when one of his research students, A. G. Davenport, wanted to test his theories by building a large boundary-layer wind tunnel (BLWT), Pugsley's lack of interest caused Davenport to move to Canada where he found the resources he needed, and quickly built an international reputation in this field. Some time later, Collar did find space for T. V. Lawson to build a small BLWT in his laboratory. The only really large space built into Queen's Building was that for the Geology Museum to

Fig. 5.5 H. M. The Queen leaving Queen's Building with the Vice-Chancellor, Sir Philip Morris, after the formal opening. Reading from left to right are H. R. H. the Duke of Edinburgh, Prof. Bruce Perry (PVC), HRH, the VC, Judge Forrest (Law Fac.) and Mr. Scully (Head Porter). The inauguration stone is shown below the photograph.

house its collection of rocks (Fig. 5.6). When Geology was relocated to the Wills Memorial Building in 1985, this space became the Civil Engineering Drawing Office, which later became the Combined Drawing Office, indicating its use by all departments.

One feature of Queen's Building which proved to be very significant for the Faculty's future was the 'pit' (Fig. 5.7), founded on rock to house

Fig. 5.6 Queen's Building. Geology Museum; now a Faculty drawing office.

Andrew Robertson's testing machine. In 1982, it was assessed by the SERC as the most suitable site in UK Engineering Faculties in which to locate an earthquake simulator, being the principal experimental facility of a national Earthquake Engineering Research Centre (EERC) which it wished to create. The EERC was a national centre in that, subject to SERC approval, researchers in any UK university could use it free of charge, and this gave the Faculty a presence in many laboratories where structural dynamics was a research activity. The award of £453,000 (with a £183,000 supplement at a later date) was the largest grant made to the University by any Research Council up to that time. Chapter 10.3 records the fact that the existence of the EERC was one of the principal reasons for the successful bid which resulted in the BLADE project.

In the MVTC, lack of space meant that workshops were a Faculty concern, but in planning Queen's Building it proved possible to have a large Faculty workshop under the aegis of Mechanical Engineering, with subsidiary workshops for each of the other Departments. This

Fig. 5.7 Queen's Building. The pit which first housed Andrew Robertson's testing machine, and then the earthquake shaking table prior to the BLADE project.

arrangement worked well, until advances in experimental techniques, particularly in electronic instrumentation, coupled with general growth, produced duplication of both facilities and manpower between these workshops. Attempts by a succession of Deans to restore a Faculty approach failed on several occasions during the ensuing years. For example, in 1985 the Engineering Board reported to Senate that it had set up a standing committee 'To review the long-term structure of the Faculty including rationalization, cross-departmental laboratories and similar considerations.' But it was not until 2005 that a full implementation was forced upon the Faculty by the BLADE development.

We shall see later in this chapter that the Faculty came under great pressure to expand almost as soon as it fully occupied Queen's Building, and to cope with this made a number of relatively small changes. These included extension of the Mechanical Engineering workshops by enclosing a passage between it and an access ramp leading to the lower courtyard; by converting the refectory into a soil mechanics laboratory for Civil Engineering; by building between columns in the courtyard for the same department's workshop, and by converting a basement cycle shed into much-needed internal storage space. Some 20 years later further space was created by demolishing a large chimney which had been a feature of the original building.

Before leaving the description of Queen's Building it is noted that Andrew Robertson's retiring gift to the Faculty was the tables and chairs in what was originally the Committee Room, but has now been renamed the Andrew Robertson Room in his honour (Fig. 5.8). In return, the Faculty's gift to him was a new motor car, together with an undertaking to maintain and repair it for him as the need arose, in the fully-equipped facility which he had himself insisted on including in the main Faculty workshop of Queen's Building for 'hands-on' use by staff and senior students. In the days when we all maintained our own machines, this facility was much in demand! He also included in his plans a room for the Vice-Chancellor, initially used by Sir Philip Morris as a private library of books which academic colleagues (as authors) had given him. As previously noted, it became the centre of University administration in 1968 during Roderick Collar's period as Vice-Chancellor[4], when a group of protesting students occupied Senate House.

5.3 The Government's Interest in Technological Education

5.3.1 Introduction
The present author's view on the difference between technology and engineering has been given earlier in this history, but important

---

[4] The Vice-Chancellor, John Harris, had died suddenly.

Fig. 5.8 The Andrew Robertson Room in Queen's Building, showing the table and chairs which he presented to the Faculty on his retirement.

bodies like the UGC and government departments persisted in using the phrase 'science and technology', with the second word to be interpreted as including engineering studies in universities; and so must it be in the early part of this chapter.

It is important to note that although the Government had supplied resources for engineering education up to this time (1950), it had not attempted to control the number of students involved, nor had it taken any significant part in the details or standards of their education. These two matters had been delegated in the mid-nineteenth century to the Professional Engineering Institutions (PEI, Appendix 2) through charters allocated to them by the Privy Council. These Institutions had the authority to admit individuals to their professions on the basis of university courses which they had themselves approved, together with, at a later stage, examinations which they conducted in order to assess the candidate's practical experience in the real engineering world. In all other European countries, governments had not allowed their PEI to take on the same important role, retaining for themselves the award of professional status in the engineering professions. It is sometimes argued that this is one of the reasons – with titles protected by law – for the often perceived difference in status between engineers in the UK and in the rest of Europe. As can be seen in Appendix 2, there was an attempt by the Government in 1971, through the recommendations of the Finniston Committee (Ref. 4) which it sponsored, to usurp the role played by the PEI, but it did not succeed.

5.3.2 *The 1950 Report of the National Advisory Council (NAC) on Education for Industry And Commerce*

This report, having the sub-title 'On the Future Development of Higher Technological Education' (Ref. 5) was relayed to universities by the UGC accompanied by a statement of its own views in 'A Note On Technology In Universities'. The Faculty cordially approved the stance taken by the UGC, which was on the general lines

> that there were technological studies based upon fundamental
> principles in appropriate groups of subjects, which depended
> on university conditions for their successful furtherance and
> these studies should continue to find a place in the universities
> of the country.

But before describing the action taken by the UGC to try to implement these views, it is necessary to record that the NAC had its own agenda, which was aimed essentially at filling the perceived gap between the education provided by the universities on the one hand, and technical colleges on the other. It proposed generous financial and building allocations to technical colleges, and promoted new courses of advanced technology in close collaboration with industry. The awards from these colleges were to be approved by a new 'Royal College of Technologists', which would, like the PEI, designate its members as Associate, Member, or Fellow, according to college and in-career attainments.

What actually emerged from this report was the creation of Colleges of Advanced Technology (CAT), one such being created in Bristol in 1958 with the involvement of the University's Vice-Chancellor, Registrar and Deans of Science and Engineering. It is of interest that a School of Pharmacy was proposed and also, in view of later developments (Chapter 7), that the University members were keen that the CAT, but not the University, should start courses leading to professional qualifications in architecture. For many, the Bristol CAT was a reincarnation of the MVTC after an interval of 50 years, and when it later raised its status to become the University of Bath, it allowed this new institution to claim that it, not the University of Bristol, had the MVTC as its progenitor.

5.3.3   The Response of the UGC to the NAC Report

It is clear from its actions during the next years that the UGC took the publication of this report as the opportunity to inform Government that the university sector in science and technology was neither large enough, nor robust enough, to meet national need. Its first success was in persuading the Government in 1953 to provide resources for the expansion of the Imperial College of Science and Technology in London. But it also alerted universities to the fact that the Government was willing to make resources available for similar developments in a small number of places in other parts of the country. Our own Chancellor, Winston Churchill, was Prime Minister at this time (Oct 1951 to April 1955) and in matters concerning science, engineering and technology – of which he professed total ignorance – it is well-known that throughout the war, and subsequently, he leant heavily on the advice of Lord Frederick Lindemann, an Oxford physicist, whose own education in the German system of Techniche Universitat, inclined him to believe that the UK would benefit by copying such a model. It seems very likely, therefore, that Lindemann (Ref. 6), as a member of Churchill's government, would have been influential in starting the process by an enlargement of Imperial College.

In this period, the allocation of resources by the UGC took place on the basis of approved quinquennial estimates, and 1953 was someway through the 1952–7 period. Nevertheless, the UGC advised that universities could amend their estimates if they so wished, and that it was willing to offer immediate additional recurrent grants specifically for developments in science and technology. The Faculty's strong card in bidding for these additional funds, in which it was partially successful, was its new, half-occupied, building. As it moved into it, the requirements for equipping new laboratories became more obvious, as did the needs of the new, but expanding, Aeronautical Department, and of its postgraduate population in all departments. But as to new initiatives, the Faculty had already said in its 1952–7 quinquennial proposals,

> The institution of any new Department before 1957 is not
> contemplated. It is realized, however, that external requests at
> a high level might arise (Agricultural engineering is one
> possibility that has been suggested). Any such request . . .
> would receive careful consideration.

But it did not pursue this possibility in response to the UGC's offer.

5.4 Conference on Engineering Education as a University Study and as a Career

All the Faculty Professors at this time were senior figures in UK engineering education and research, being as aware as the UGC of the need to expand the national effort in these areas, and this was one reason why they organised a two-day 'Conference on Engineering Education' in April 1955. But they also had their own Faculty's needs very much in mind for the following reason. From its beginning in 1909 until 1951, the Faculty was proud to report the number of students who came from a colonial or dominion country, and the University had a Colonial Students Committee[5] to ensure that their special needs were provided; but by far the greatest number of students came either from the Bristol area, or from neighbouring counties. Like the other 'civic' universities[6] in the UK, it catered primarily for local students, and could not be regarded as a national university, let alone having European or international recognition. To begin to obtain such a status was a second reason for this conference.

It was held at Wills Hall, where Headmasters, Directors of Education, members of education authorities and industrialists were guests of the University. Its aims were to explain and emphasise that engineering offered an education in its own right, leading to a useful and valuable career in many professions. It also emphasised that there was a great and growing need for engineering graduates, and that the new building in Bristol offered facilities unsurpassed elsewhere in the UK.

The central position of mathematics in modern engineering studies was recognised, and it was pointed out that in the Bristol Faculty there were two alternative honours degree courses which catered for

---

[5] It became the Overseas Students Committee in 1956.
[6] The terms 'red-brick' and 'plate-glass' were used later to describe two groups of new universities.

different levels of mathematical ability, in addition to the ordinary degree course. Also, that the intermediate course, which introduced students to the basic elements of engineering science, was available for those who had studied 'arts' subjects at school.

The Vice-Chancellor, Sir Philip Morris, chaired the conference, and Sir Harold Roxbee Cox (later Lord Kings Norton and Pugsley's wartime colleague) initiated a discussion on the country's need for broadly educated and trained engineers.

Besides informing secondary schools about engineering studies in the University, the conference highlighted the lack of any formal link between schools in the area and the University. This was remedied in 1960 by Jack Diamond, a local MP who, following a suggestion by the Professor of Spanish, Jack Metford, gave the University £5000 for the establishment of a 'University and Schools Committee'. Its first chairman was Bruce Perry, Professor of Medicine, and the early meetings included such topics as 'Science Teaching in the Schools and in the University,' and 'The Problems of Transition from School to University.'

5.5 Continuing Pressure for Expansion, 1956–60

In 1956 the UGC was required by the Conservative Government of Harold Macmillan to consider the situation arising from the postwar increase in the birthrate, estimated to be an increase of 40 per cent on the current number reaching university age in 1964, and still 15 per cent above the current number in 1971. In turn, the UGC attempted to persuade universities that it was 'a matter of social justice' to make provision for these students. In reaching its target of 176,000 by 1970, the UGC envisaged the creation of new universities, but supposed that in the short term these could only make marginal contributions to the foreseen problem.

The UGC specifically targeted smaller and medium-sized universities, and since Bristol fell into the second category, it was asked to increase from 3500 to 5000 students during the period of peak demand. Whilst this was regarded as unwelcome in many respects, the University accepted it, with the proviso that the UGC provided the necessary funds to build residential accommodation 'in greater proportion to the increase in the number of students'. At this time, the only halls of residence for men were Wills Hall, Burwalls and Churchill Hall, whilst for women there was Clifton Hill House and Manor Hall.

The Faculty was not easy with this acceptance of expansion, because the Government's wish was that at least two-thirds of the additional students should be in science and technology. In a long and cogently argued report to Senate, its Dean, Prof. Collar, pointed out that well into the 1957–62 quinquennium the Faculty would be concerned with the proper utilisation of its new building (not called Queen's Building until the end of 1958), the assimilation of a new Chair in Theoretical Mechanics, and 22 other academic appointments, which effectively meant a doubling of the work of the Faculty in half a decade. However, pressure must have been applied by the Vice-Chancellor to force the Faculty to play its part in the expansion which the University had agreed, because it did increase its total undergraduate number from 256 in 1955 to 325 in 1957. In that same year the UGC wrote again to the University with the information that it would now be regarding the 5000 student figure, not as a 'peak', but as a 'plateau' figure.

As usual, the UGC had not been well served by the inability of Government departments to produce accurate statistics, and so, in February 1960, it had to make further requests for expansion. In writing to the University it made the following comment:

> Not only will the 'bulge' in the later sixties be larger than
> previously expected, but the succeeding 'plateau' will be
> shortlived and will be followed by a still larger expansion (due
> to 'trend') in the seventies. The future number of potential
> students will therefore be considerably greater than expected.

The UGC explained that what was meant by the word 'trend', was the actual increase in the proportion of the 18-year old cohort who were expressing a wish to attend university, coupling this with an observation that in their view, there did not appear to be any lowering in the quality of the applicants. This last comment was probably in response to an unease in many quarters at this time that 'more means worse', a phrase much used in newspaper and journal articles by the author Kingsley Amis. His later novel *Lucky Jim*, made into a very popular film, characterised a hapless lecturer in History in a provincial university who revolted against the pompous posturings of his senior colleagues. The fact that Amis had himself been a lecturer in English at a provincial university was generally assumed to have given him material for his novel, which portrayed both staff and students in a very bad light!

The Committee of Deans was willing to accede to the UGC request, recommending expansion to 6500 students, subject only to the provision of additional accommodation. But the Dean of Engineering, John Morrison, was likely to have expressed a minority view in this Committee because the Engineering Professors made the following statement:

> We think that to undertake yet another expansion would seriously militate against the academic health of the University. We believe that the above general view applies particularly in the Faculty of Engineering, where the senior staff has already been heavily preoccupied with expansion matters for over ten years, and where a continuance of this process is likely specially to damage its research work, the need for extension of which was one of the main aims of the new building.

They went on to suggest that it would be appropriate to consider further expansion at the end of the next quinquennium, that is, in 1967. It is clear, however, that the University continued to apply pressure on the Faculty, in particular by publishing estimates of an intake of students to the Faculty and an aggregate population which the Faculty had not authorised. But in bowing to this pressure the Faculty saw an opportunity for its own advancement. It reported to Senate in June 1960 that it had undertaken a study of a more effective utilisation of its space in Queen's Building, and subject to obtaining additional academic and technical staff of adequate quality, and to the provision elsewhere of accommodation for the Department of Mathematics, it could increase its annual intake of undergraduates from 120 to 155, resulting in a student population of over 450. The largest increase would be in Mechanical Engineering, with smaller increases in Electrical and Civil, but none in Aeronautical, this distribution being attributable to the current state of competition for admission, and the demand for graduates. The Faculty had little difficulty in recruitment because the ratio of applications to places during 1957–60 varied only between 12.5 and 13.0 – which was at the higher end of the University's range – with few entrants having less than three good A-level passes. In addition to this change of attitude on undergraduate numbers, the Faculty also contemplated cooperation with the Science Faculty in the creation of a Department of Chemical Engineering.

But during the next few months the Faculty yet again reconsidered its position, and in May 1961 made another lengthy report to Senate in which it offered the alternatives for expansion of *either* the 120 to 155 intake of undergraduates, *or* an expansion in postgraduate numbers from 16 in 1959 to 45 in 1964. It clearly preferred the latter, but, in a conciliatory tone, expressed a willingness to revert to its original proposal '. . . in the light of the national interest and the circumstances of other similar departments elsewhere.' It had also become aware of negotiations then taking place for the absorption of the West of England Academy School of Architecture into the University, noting that such a development would require additional Faculty resources with which to offer '. . . basic courses in fundamental engineering disciplines'. As described in Chapter 7, Prof. Pugsley, then Pro-Vice-Chancellor, was the driving force behind these negotiations, which envisaged a School of Architecture having a strong base in engineering science, requiring considerable input from the Faculty. The revised details of the Faculty's requirements for the 1962–7 quinquennium now included a Chair in Engineering Thermodynamics (occupied by G. F. C. Rogers in 1964), and 'About seven calculating machines: total cost about £800 (irrespective of the installation of a large computer)'. The computer reference here is to an application recently made by the University to the UGC to purchase such a machine, and the establishment in 1961 of a Committee on Computer Science '. . . to consider proposals which have been put to the Vice-Chancellor for the development of theoretical studies relevant to the work done by computers'; its chairman being the Pro-Vice-Chancellor, Prof. Pugsley.

As a postscript to this havering attitude concerning expansion of undergraduate numbers, the Faculty persisted in pointing out that expansion of the research activity had been a major part of its case for Queen's Building and that the present objective was to achieve research excellence in all Departments. Those in the Faculty today will smile ruefully at their comment:

> It is extremely difficult for a member of staff to find time for research work which should be regarded as of comparable importance with teaching and as an essential complement to it, but which all too often has to be neglected.

At the time, their student/staff ratio was ten, and despite its forebodings, during the next 15 years it achieved international excellence in many of its research activities.

5.6 The Robbins Report on the Future of Higher Education in the UK

Following the piecemeal approach to university expansion just described, Prof. Lord Robbins, an economist by profession, was appointed in 1961 by the Conservative Prime Minister, Sir Alec Douglas Home, to be chairman of a committee to consider the future of higher education in the UK. One of its members was our own Vice-Chancellor, Sir Philip Morris.

Robbins' report (Ref. 7), presented to Parliament in 1963, called for a really major expansion in student numbers from 216,000 in 1962–3 to 560,000 in 1980–1, within which it is important to note for our purposes that there was to be a significant component of what he called 'Scientific and Technological Education and Research'. There were to be five special institutions created in these areas '. . . comparable in size and standing and in advanced research to the great technological institutions of the USA and the Continent.' He considered that the basis for three of these existed already, at Imperial College, the University of Manchester Institute of Science and Technology (UMIST), and the Glasgow Royal College of Science and Technology – later to become the University of Strathclyde. The recently created Colleges of Advanced Technology were to be given the status of Technological Universities, and the volume of postgraduate work in science and technology was to be considerably increased. It is noted that the CATs thwarted Robbins' intentions by becoming multi-faculty universities of the more conventional type; for example, the Bristol CAT became Bath University.

To oversee these, and many other changes, a Grants Commission was to replace the UGC, and a Government Minister for Arts and Science was to be responsible for the greatly increased number of universities, as well as for the Research Councils.

The Government and the universities responded positively to the Robbins Report, urging the UGC to accelerate any existing proposals for expansion. For our University, it suggested an increase in student numbers by 1967 beyond the existing target of 5200, so as to reach 6000 in that session. The University accepted this on condition that finance would be provided for a list of 12 major building projects, which included five new halls of residence and a substantial increase in the minor works allocation so that more student residences could be provided by modifications to existing buildings. Not surprisingly, no additional accommodation was envisaged for the Faculty, but it can be noted that in 1962 the University administrative staff, from the Vice-Chancellor downwards, moved from what, in 1964 was to be officially named as the 'Wills Memorial Building (WMB)', to the new Senate House on Woodland Road. As an indirect consequence of this, some 20 years later, Geology moved from Queen's Building to the basement of the WMB.

The Robbins proposals for expansion in student numbers followed those made earlier, by weighting science and technology in a two-to-one ratio with arts, but although this may have been desirable nationally, there were neither sticks nor carrots to persuade schoolchildren to take up such disciplines at university. In the out-turn therefore, the real expansion at this period, in Bristol and elsewhere, was weighted in the opposite direction, causing Prof. Powell, our Nobel Prize winning physicist, to draw attention in Senate

> to the increasing proportion of Arts students in the University arising from the increased admissions to that faculty over the last two years. It had been the original intention to increase the proportion of students in the Faculties of Science and Engineering in correspondence with the wishes of the UGC but the present trend showed that the opposite was happening.'

Powell's comment was repeated many times by the Science Faculty during the next years, during which it found increasing difficulty in recruiting adequately qualified students, due, in part, to the unenthusiastic teaching of science subjects in schools.

5.7 Implementation of the Robbins Recommendations

5.7.1 Student Numbers
No government can commit either itself, or its successors, to the 20-year expansion proposed by Robbins. Instead, in sympathy with the general intentions, it provided the necessary financial resources in the 1962–7 quinquennium for a modest increase to 225,000 students,

emphasising that undergraduate numbers were a genuine priority. But in distributing this number to individual universities, the UGC encouraged them to admit more than the number of students which it had authorised in its separate allocation letter, '. . . if by internal economics, increased productivity or *any other means* (my italics) it thinks it can rightly do so.' It also took cognisance of reality in the matter of student preferences in the subjects which they wished to study, by reversing the balance between arts on the one hand, and science and technology on the other – it now accepted that the major increase must be in the arts-based subjects.

5.7.2   Unit Costs

It was no surprise to anyone involved in university administration, at any level, that a consequence of increasing government involvement meant greater financial control. This emerged in 1965 by the UGC requiring financial returns from universities on a cost analysis basis under the main headings of undergraduate teaching, postgraduate courses and research, and the research of teaching and research staff. The UGC attempted to reassure universities that they would indeed expect to find some differences in unit costs between one university and another, but in a scarcely veiled threat, they actually said that

> Relative costs have, therefore, been one of the factors to which the Committee have had regard to in making their allocation, and they have made some adjustments where there *seemed to them* (my italics) to be over- or under-financing of individual universities.

5.7.3   Postgraduate Numbers – The 'Brain Drain'

Looking back, it is scarcely credible that the UGC could have agreed to implement the Robbins Report by asking universities to reduce the number of postgraduate students and, in doing so, without making any reference to individual disciplines. Their reasons for taking this extraordinary action now appear contrived, and include

> . . . there is uneasiness that the rise in the proportion of graduates who stay in the university for postgraduate studies, rather than moving into teaching or the outside world, is greater than the country can afford at present.

Whether the members of the Robbins Committee were competent to make an accurate numerical estimate of the national need for post-graduate students is in doubt, which makes questionable the UGC statement that any such estimate had already been exceeded. That Robbins, via the UGC, could have called for a reduction in postgraduate numbers is all the more surprising because only three years later, in November 1966, the Labour government asked Dr. F. E. Jones, managing director of Mullard Ltd, to chair a committee

> To study the international migration of qualified engineers,
> technologists and scientists as it affects the United Kingdom;
> to identify both the advantages and disadvantages, and to
> make recommendations accordingly.

This topic of concern came to be referred to as 'The Brain Drain' (Ref. 8), the largest exodus being to North America. It is pleasing to note that whoever drafted the above Terms of Reference of the Jones Committee, actually understood the difference between engineers and technologists. In its October 1967 Report, whilst concluding that there was a serious brain drain in those three categories of manpower, it made the recurring criticism of Government, that its statistics were inadequate for the purpose of making definite conclusions, and that its responsibilities in this important area were divided between too many departments. Three of the 22 recommendations were directed at universities. For instance, one said that they should recognise work in industrial laboratories for higher degrees, that they should consider linking or combining faculties of engineering, technology and science, and that they should reverse the tendency to train scientists towards academic achievement as an end in itself and should direct the emphasis more towards the manufacturing industry. In our Faculty it took some years before its Regulations allowed the first recommendation to be accepted, but still requiring for such researchers that a minimum of one year should be spent in the University. The second recommendation did not apply with any force to Bristol because there was already much common activity between the Faculties. As to the third recommendation, although restricted to scientists, the less said the better!

5.7.4   Collaboration with Industry

Here also, the UGC was not at ease in the advice which it gave to universities regarding the Robbins Report. It accepted that a university has other objectives besides providing industry with well educated employees, but goes on to say

> ... it would be valuable if universities collectively made a
> further and deliberate effort to gear a larger part of their
> output to the economic and industrial needs of the nation, for
> few things could be more vital to the national economy than
> the proper deployment of highly qualified scientific
> manpower and the application of research to the solution of
> current technological and economic problems.

How this could be consistent with a reduction in postgraduate numbers is not made clear. Certainly, the lack of 'joined-up' government is not entirely a modern phenomenon!

5.8 The Faculty Plan for the 1967–72 Quinquennium

Although the Faculty Professors agreed with the UGC that there was no national pressure from schools to increase the number of undergraduate places in engineering, it reflected its own recent experience by arguing that the demand for such places at Bristol was likely to increase. They certainly did not agree with the UGC about postgraduate numbers, again pressing the University to find accommodation for Geology and Mathematics so as to free 30,000 sq ft in Queen's Building for increased research activities. Whether as an alternative to this, or in addition to it, is not clear, but a building extension was proposed below Queen's Building on the Woodland Road frontage.

The specific proposals for development were the MSc course in industrial thermodynamics, due to start on October 1966 with expansion during the quinquennium, and the establishment of a Department and Chair in Electronic Engineering. The Departments of Civil and Aeronautical Engineering were to develop industrial aerodynamics as a joint activity, and the second of these two Departments was to develop the use of computers in aircraft design. If all these plans came to fruition, student numbers would increase by 160 – from 364 in 1962. Academic and technical staff would need to be increased by 15 and 16, respectively over the five-year period. In fact the expansion was much slower, a total student number of 532 (Table A4.5) not being achieved until 1981.

The principal reason why Electronic Engineering only now appeared on the scene in Bristol was the reluctance of Prof. Rawcliffe to lose any of the autocratic control which he exercised over his Department. In 1967 his professorial colleagues had persuaded him that this subject was an essential development, and in 1968 Senate had established a committee charged with making an appointment, with Rawcliffe as convenor, and the present author as a member. But one year later, Senate asked why no progress had been made, prodding Rawcliffe into action; even so, it was not until the middle of 1970 that he found what to him was an acceptable occupant of the Chair. In fact, Kenneth Sander, Senior Tutor at Trinity College Cambridge, proved to be an excellent choice for the Faculty – a scholar and teacher of the first rank, but with no ambition to usurp Rawcliffe's role in controlling what had become the Department of Electrical and Electronic Engineering. Sander's own research interests were such that he was content to occupy a small corner of the basement bicycle shed, and so did not intrude on Rawcliffe's extensive domain.

Rawcliffe's Royal Society Biographical Memoir, written by two of his Bristol colleagues who knew him well, comments that '. . . he had a vivid turn of phrase and a conceit so innocent that no one could take offence at it.' This was well expressed, because his conceit was such a part of the man, that more often than not it caused amusement, even laughter, as for example at a Senate meeting when he asked, as Dean, 'That clear information be given on the appropriate dress for wives and women members of staff.' Here, he was objecting to trousers! Perhaps less amusing, expressed in more private company, was his view that 'One of the grounds on which a man ought to be able to divorce his wife was intellectual incompatibility.'

5.9 The Hale Report on University Teaching Methods

The committee chaired by Sir Edmund Hale was established in 1961 (Ref. 9). It was another illustration of the Government's intentions, having taken over the funding of universities, to control the details of their activities. It was, indeed, the beginning of time-consuming periodic investigations of teaching practices of each department by

Government-appointed bodies in the name of 'quality assurance'. This term was to be applied to the academic staff themselves as well as the courses which they gave, leading to the requirement in all universities that they should attend 'induction courses', so that their approaches to these responsibilities could be monitored and approved!

It was illuminating, however, for the Faculty to find that it could give positive answers to the questions on such matters as its allocation of a personal tutor to each student, the use of vacations for field courses and practical experience, and the assessment of a student's progress through written and oral examinations. But one issue raised by the Hale Committee was the possibility of a course structure which made it possible for a student to move from one department, or even faculty, to another. Engineering had always restricted this possibility to first-year students transferring only within the Faculty, and even then, warning them to study for the transition during the first long vacation. But 40 years on, as we shall see (Chap. 8.11), it was required to embark on a lengthy process of allocating 'credit points' to each course in the curriculum, so that they could be carried between departments, faculties, and even universities both at home and abroad, as contributions towards a degree.

Whether the implementation of the Hale proposals have actually improved the quality of what students take from their university experience is open to question, and would depend on the innate qualities of the individual. Our Faculty has always been able to recruit high calibre students and staff, the latter often having acceptable idiosyncracies which actually added to their effectiveness as teachers. It took the view that the real requirement in a university environment is that students and staff join together in a learning experience informed by the research and professional experience of the teacher. What was not required was an imposed format of teaching style.

5.10 The Dainton Report into the Flow of Candidates in Science and Technology into Higher Education

Dr F. S. Dainton was the Vice-Chancellor of Nottingham University and later the Chairman of the UGC. His small committee of five was set up in February 1965, almost two years before the Jones (Brain Drain) Committee and included two members of that committee, Lord Willis Jackson, FRS of Burnley, an eminent electrical engineer, and Prof. L.

Rosenhead, Head of Applied Mathematics at Liverpool. In the matter of distinguishing between engineering and technology their influence was significant, because although Dainton's terms of reference does not do so, his recommendations do make the distinction in all instances where it is relevant (Ref. 10). Thus, in the two recommendations concerning universities specifically, it first suggests that they

> reconsider entry requirements with a view to encouraging a broad span of studies in the sixth form and to increasing the flow of candidates into *science, engineering and technology* (my italics).

Its second suggestion was that

> Universities should further experiment with new courses in science, engineering and technology designed to attract to these disciplines able entrants who are not already committed to these fields of study.

In addressing schools, Dainton recommended that all pupils should study mathematics until they leave school, and linking with the first of the above suggestions to universities, it can be said that a satisfactory performance in mathematics has always been the only prescribed scholastic requirement for entry to the Faculty, coupled with an equally satisfactory performance in at least two other subjects of their choice.

For the second of the Dainton recommendations to universities, reference back to section 5.4 of this chapter will indicate that the one-year intermediate course was offered to such school-leavers wishing to transfer from 'arts' to engineering, but pressure from many sources, including the wish to strengthen its research, caused the Faculty to abandon this opportunity in 1966.

One change which might have helped to implement Dainton's second recommendation was a 1974 proposal to change the sixth form curriculum and examination system from A-levels, three of which were normally taken, to a system in which it was intended that five subjects would be studied, three at Normal (N) level and two at Further (F) level. The N-levels would require about half the teaching time given to A-levels, and F-levels about three-quarters. The aim of these proposals was to defer the need for choice by schoolchildren of their intended further education subject, and the avoidance of early specialisation. The proposals, which were accompanied by draft syllabuses in many cases, had the support of the Standing Conference on University Entrance (SCUE), but when referred to the University for comment, no Faculty or Department welcomed them. The Faculty, together with Science and Medical Faculties, took the view that the N-levels would not be a sufficient basis for a three-year (five in medicine) honours degree, and in the absence of longer degree courses these Faculties would insist on four F-levels in relevant subjects, because it was clear from the proposed syllabuses that only the results at this level would indicate the applicant's capacity for advanced study.

The general critical response which the N-F system produced led to its disappearance from the educational scene, but there remained an undercurrent of dissatisfaction with what was seen in the A-level part of the General Certificate of Secondary Education (GCSE) to be a constraint against a broader education at sixth-form level, with many critics pointing to the advantage of the Continental Baccalaureate.

This dissatisfaction in some quarters with the A-level system of admission to universities surfaced again in 1983, with the idea of broadening sixth-form studies by the introduction of AS-levels, to be taken probably in the first sixth-form year, with A-levels to be taken in the second year. To some extent the idea came from the 'Higher Education in the 1990s' Government Green Paper (Ref. 11) which is discussed in detail in Chap. 8.7. It wished to see more young people studying mathematics and sciences into the sixth form as part of a general education in business and commerce. The Faculty decided that there would be no change to its requirement of A-level passes in mathematics and in physics (or engineering science as offered by various examining boards), but that there would be no difficulty in accepting two AS-levels as being equivalent to the third A-level, so long as these were in such subjects as indicated a broad sixth-form education.

One practical effect of the introduction of the AS-level was to double the numerical values attached to the A-level grades, which became 10, 8, 4, 2, allowing the AS-level grades to be allocated 5, 4, 3, 2, 1.

5.11 Proposals for the 1972–77 Quinquennium

Preparation for this period was a difficult exercise for the Faculty. Of the seven Chairs in the five Departments, counting the new Chair in

Electronic Engineering[7] which was occupied in 1971, four were due to be vacated during 1971–5, and these carried the Headships of the Departments of Mechanical, Electrical and Aeronautical Engineering, together with that of Theoretical Mechanics. It was inconceivable that these Chairs would not be filled at the earliest possible moment, and such a large-scale turnover of senior posts would generate substantial additional needs for staff, equipment, accommodation and budgets.

The high ratio of applications to places in the Faculty, and the greater interest in electronics, prompted it to argue that in the national interest, the University should allow it to increase its undergraduate intake from 130 to 150 a year, and postgraduates from 50 to 80, each Department taking a share of these increases. In the quickly developing Aeronautical Department new activities in industrial aerodynamics, structures and high speed/low density flows would require three new academic posts, whilst Civil Engineering would need one extra post to cover its increased teaching contributions to Architecture (Chap. 7). All Departments would need additional technical staff and accommodation, with regard to which the Faculty again proposed that Geology and Mathematics should be found alternative accommodation, the latter in an adjacent old house in the precinct called Tower View.

The 1972–7 quinquennium was the last time that the UGC was able to plan in this way. The increasing rate of national inflation caused this system of funding to be replaced by a year-on-year model which made management of universities very difficult.

5.12 The Faculty and the European Economic Community (EEC) – the Common Market

Even had the Faculty known that the Prime Minister, Edward Heath would sign the Accession Treaty to the Common Market in January 1972, the 1972–7 quinquennial proposals would have been only slightly modified, because few understood its longer-term significance for education in the UK, particularly for those departments in universities which were concerned, as Engineering was, with preparation of students intending to follow careers in a profession.

---

[7] From 1971 the Electrical Engineering Department became the Electrical and Electronic Department, but for brevity in this history, its original title will be used, unless the distinction is relevant.

The actual decision on accession had been voted through in the House of Commons in October 1971 and the Faculty had almost immediately discussed some of the ways in which the education of engineers might be affected. It decided that no fundamental changes were required in the courses offered by the Faculty, but that consideration should be given to requiring candidates for admission to have passed in one European language at O-level, and that all students should be actively encouraged to use the University language laboratory. It will be recalled that in the early days of the Faculty, courses in French and German had been part of the curriculum, but this requirement had been allowed to lapse, because most students already had such an O-level. The Faculty further suggested that lectures should be given to undergraduates on the structure and policies of the EEC, and that the contents of the Engineer in Society course might need to be enlarged to cover relevant topics. From the immediate post-war period until 1969 engineering students had been required to take examinable courses in English and Economics given by academic staff from these University Departments, but at this date Law and Politics had been added to the syllabus of this course. In view of an earlier remark about lectures given by one department to students in another being regarded as a 'chore', it should be recorded that for three of these subjects it was the Senior Professor who chose to give the lectures to engineering students, because, they said, it was a refreshing experience to do so!

As we shall see in Chapter 9, accession to the EEC, later to become EC – European Community – did make major changes to the Faculty, principally on the need for agreement on equivalence of degree standards and professional qualifications, which, in turn, and with financial support from EC educational programmes, generated flows of students between the Member States. Many of these issues were addressed by the CVCP in its report 'The European Community and Higher Education' (Ref. 12) particularly that of professional qualifications arising from the mobility of labour.

5.13 The Beginning of Financial Stringency

Early in 1973 the UGC were still planning for a 1981–2 full-time student cohort of 375,000, but towards the end of that year it had to tell universities that government cuts in public expenditure meant a cut in grant allocations in 1974–5 and beyond. The effective cut for Bristol amounted to 15 per cent, requiring that all vacant posts be frozen, and no new posts allowed. By the end of 1974 the rate of inflation in the UK was expected to rise to around 20 per cent, which meant that no alleviation of this financial stringency could be foreseen during the 1972–7 quinquennium, with little hope of the Faculty achieving its development objectives. But inflation did not quite reach this level, making it possible for the government to make supplementary grants, and by some manipulation of its statistics, was able to reduce the expected number of full-time UK university students by 1981 from 395,000 to 320,000. With counter measures to reduce the 'Brain Drain' still in place, the Faculty rightly assumed that this reduction did not apply to engineering students. In Bristol, the shortfall of UGC funds for building projects meant that the conversion of the Wills Memorial Building for use by the Arts Faculty would not be completed by the end of the quinquennium (1977), with a consequent reduction from 2934 to 2534 in the number of arts-based undergraduates which the University had undertaken to accept in its 1972–7 proposals to the UGC. On the other hand, the number of science and engineering undergraduates remained the same at 3018, with engineering postgraduates increasing from 44 to 53 by 1973.

There was no relief in financial stringency during the latter half of the 1970s, and the screw was then tightened with increasing vigour in the early 1980s until it became necessary for the University to choose between distributed misery or the closure of one, or more of its Departments. This difficult period for the Faculty will be discussed in the next chapter.

EXTRACT FROM UGC ANNUAL SURVEY 1976/77 EXTENDED TO 1978/79

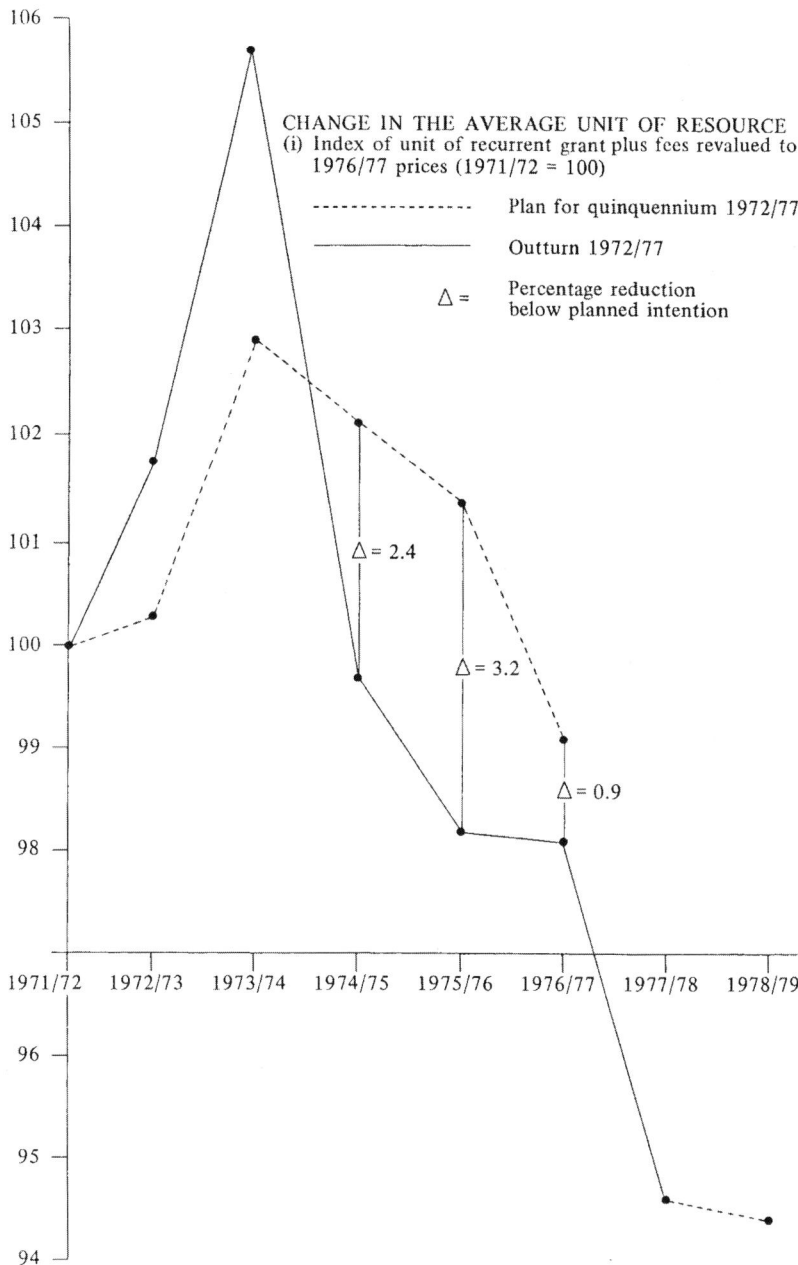

CHANGE IN THE AVERAGE UNIT OF RESOURCE
(i) Index of unit of recurrent grant plus fees revalued to 1976/77 prices (1971/72 = 100)

- - - - - - - - - - - - - - -   Plan for quinquennium 1972/77

────────────   Outturn 1972/77

Δ =   Percentage reduction below planned intention

Fig. 6.1 Changes in the unit of resource during 1971–8.

1   it indicates a misunderstanding of the role of universities in the education of chartered engineers;

2   it is a divisive proposal and, what is worse, it will be divisive without achieving its intended purpose, which presumably is to get more (and/or better) boys and girls to study engineering and science;

3   the problem of getting more good people into wealth-producing industry is complex and will not be solved by this method.

But, unlike engineering in some other universities, our Faculty experienced no difficulty in attracting an increased number of good candidates (Table A4.5), and its main requirement for the quinquennium was the enhancement of computing facilities, particularly graphics terminals linked to the University computer, and a Faculty programmer 'to assist with the teaching which uses computers'. It had also observed the lavish accommodation provided in the newer universities and polytechnics, giving strength to its renewed request that Geology and Mathematics be found accommodation outside Queen's Building.

In line with government and UGC wishes, undergraduate numbers had recently increased, and in 1976 stood at 450. A stable future population of 480 was envisaged, including students in the new degree course of Engineering Mathematics. This was considerably below the 580 undergraduates which the UGC capacity figures for Queen's Building had stated, but it is possible that their officials were unaware that Geology and part of Mathematics were also being housed there. The number of postgraduate students was 46, also much lower than the UGC requirement of 17 per cent of the total number of students, but nearly all these were PhD candidates, the Faculty as a whole not being in favour of higher-degree taught courses – taking the view that its prime function was excellence in both undergraduate teaching and in research.

The impact of the UK joining the EEC in 1972 was under consideration by the Faculty, particularly the fact that EEC regulations required a minimum of 4 years of university education as a preliminary to qualification as a Category-A engineer, but at this time it could only wait upon the results of initiatives which the Professional Engineering Institutions were taking at government level.

In Civil Engineering the quinquennial estimates drew attention to its historical concentration on structural engineering, the balancing of which required new academic staff in water engineering and soil mechanics; a Professorial post in one of these was requested. Responding to the Robbins proposals (Chap. 5.6) Mechanical Engineering had already increased its undergraduate intake from 30 to 45 without additional resources, but was willing to increase its number of postgraduates if university-funded research posts could be created. Electrical Engineering had new Professors in both its Electrical and Electronic sections, and would require the quinquennium to consolidate new courses and to build up their research activities. The Aeronautical Department also had a new Head whose interests in spacecraft, rather than more conventional aircraft, would take time to develop, and this, together with expansion in industrial aerodynamics research, would be the Department's programme for 1977–82.

6.4 The Cuts Bite Deeper

Inflation was such that for the session 1976 the UGC were forced to make a provisional financial allocation to universities, followed later by a supplement to cover pay increases, but it was coupled with the dismal information that the resources by way of grant and student fees would be four per cent less in 1977 than in 1976. However, because some inflationary increases in university expenditure were unavoidable, this meant that the cut would be six per cent in items such as salaries, the total of which could only be varied if the damage caused was acceptable, for example by freezing posts. There were hints that the Government intended to raise student fees, but since these amounted to only about 28 per cent of the University's income, such an increase would not make a meaningful difference; even so, it would only add to the difficulties if numbers were to fall below the targets for the undergraduate intake. With regard to fees, for most students these were paid to the University by the student's Local Education Authority, which, in turn, was reimbursed by Government to the extent of around 90 per cent, the remainder coming from local rates

The significance of a four per cent cut was that in Bristol it would mean the loss of between 30 and 40 posts, and although it was possible that these could come from retirements and resignations, such adventitious losses were not likely to be in the University's best interest. To safeguard these interests, compulsory redundancy would have to be considered, but a total closure of any department was not contemplated.

In March 1977 the UGC was informed by Government that its funding of universities for 1977 would be one per cent below that for the previous year, even though a growth of five per cent in student numbers would still be expected. The CVCP again regarded the Government's figures as seriously inaccurate, concluding that the shortfall in funding would be closer to three per cent, which could be interpreted as a loss of 25–30 academic posts in Bristol, offset slightly by an increase in tuition fees which the government had imposed – to £650 p.a. for undergraduates and £750 for postgraduates – an action which caused a student sit-in in Senate House!

The Government's calculation of the demand for full-time student places varied periodically; as has been seen, it had at one time been 395,000, then 320,000, followed by a further reduction to 291,000, but in January 1978 it raised it again to 310,000. Its argument for doing so was that when universities had been asked in 1977 for their expansion plans to the 1980 session, the aggregate number which they were willing to accept had been 305,000, which was virtually what the Government was now proposing! For Bristol this meant that the target figure for 1981 was now 7600 full-time students, and although the University's view on its capacity was at variance with that of the UGC, it accepted this target figure and asked faculties to draw up plans to achieve it. Regarding the balance between arts-based and science-based students, the University already matched the UGC figure for the national average and so anticipated no change, but it strongly opposed the required reduction of three per cent in overseas student numbers. In response to this opposition by the great majority of UK universities, the UGC issued a clear warning '. . . that if the proposed targets were not met then there is a distinct possibility that the provisional grant for 1981 would be reduced.'

Although the Bristol target figure for 1981 of 7600 was delayed by the UGC for one year, its intake targets for 1977 were not achieved, putting the onus on doing so in the sessions 1980 and 1981. Failure to do so would lead to an income deficiency of around £300,000 per annum, and a consequent prospect of enforced redundancy, in respect of which it had to be accepted that the average student/staff ratios for all faculties in Bristol – ranging between 8.63 and 7.56 for the period 1971–8 – were slightly lower than the national average.

By the middle of 1979 the UGC had been forced, yet again, to change its intake target for 1981, requiring universities 'to so arrange their admissions procedures that the number of home undergraduates being admitted in October 1980 can if necessary be restricted to 94 per cent of the number being admitted in October 1979.' At the same time, it told universities that '. . . the recurrent grant will be held for a number of years at no higher level in real terms than that in 1979.' The Vice-Chancellor, now Sir Alec Merrison[1], was at this time Chairman of the CVCP, which gave him the opportunity of direct contact with the Conservative Minister for Education and Science, Mark Carlisle, during which he was able to explain, from a national viewpoint, the folly of the Government's actions, particularly the greatly reduced funding contribution for overseas students. Sir Alec calculated that universities would need to charge these students £3500 per annum to recover the loss which the Government was imposing, and pointed out that this would cause foreign students to seek universities in other European countries, North America or the Soviet Union where fees were very much lower, and that this would be an economic, political and educational tragedy for the UK. The only known outcome of Sir Alec's efforts was the notification by the UGC that the *minimum* fees for overseas students in 1981–2 were to be £2500 for arts courses, £3600 for science and £6000 for clinical courses.

6.5 Selective Cuts or Distributed Misery?

It is unlikely that any chairman of the UGC, or its successor organisations, has had to write such a morale-sapping letter to the Vice-Chancellors of UK universities as Edward Parkes was required to do in May 1981. It advised them that the total loss in their government grant by 1983, compared to 1979, would be 'at least 11 per cent and possibly significantly more'. The pain was not to be spread equally, either between institutions or between fields of study. But whereas it contemplated reducing the *range* of subjects taught at some universities, which effectively meant the closure of departments, it did not expect the closure of any whole university. What it most certainly did expect was redundancy of all grades of university staff, together with early retirements as an aid to restructuring, and for this purpose it had set aside

---

[1] The knighthood was well deserved. He had undertaken many responsibilities outside the University. These included, – all as Chairman – an investigation into the National Health Service, the box-bridge failure enquiry, and the CERN nuclear laboratory in Geneva.

£20m, access to which required an acceptable restructuring plan from each university before the end of January 1982.

To understand the effect which the UGC letter had on the Faculty, it is noted that at its meeting in October 1981, Senate favoured selective cuts rather than distributed misery, and in doing so gave the Vice-Chancellor the difficult task of formulating detailed proposals to achieve this. In a parallel activity, all Departments were required to examine their own possible contributions to the necessary economy, making a report to what was referred to at this time as Committee I, whose terms of reference were

> To examine the 1983–84 plans of Heads of Departments for
> their financial and academic reliability, and to make proposals
> to the Committee of Deans and the Estimates Advisory
> Committee.

Here, the 'plans' were statements of savings in total expenditure through loss of staff, postponement of agreed developments, and realistic ideas for generating new income. The word 'Department' covered all spending units in the University; for example, in the Developments Committee it was assumed that there would be a reduction in student numbers, allowing it to raise capital by selling some of the stock of student residences, and the Bursar's Department did begin the process of selling a strip of land at Coombe Dingle. In the Student's Union, there were plans to house Student Services, but the cost of doing so was prohibitive.

But the really difficult task was given to another group, Committee II, whose remit was

> Where Committee I reports that compulsory redundancy is
> unavoidable, then to select the posts of individual members of
> the academic staff for such redundancy and to recommend
> accordingly to Senate and Council.

The Vice-Chancellor's plan, based upon the very detailed type of analysis which one would expect from a Fellow of the Royal Society, compared Bristol's departmental unit costs with national averages. It led him to a proposal to require about one-half of the Faculty of Education to move to a self-funding basis, and to close completely the Department of Architecture and the small research unit for Comparative Animal Respiration altogether. In making these proposals he would have had some idea of the voluntary emasculations which the various departments would be able to offer through Committee I, and, in the case of Architecture, the UGC's published advice that there was an over-provision of Departments of Architecture in the UK, some of which were too small to be efficient, and that such departments as were geographically close could increase efficiency by merging. It is known[2] that such a merger between the Departments at Bristol and Bath had been encouraged by the UGC, but, as Chapter 7 describes in detail, the two parties were not willing to even contemplate such action!

For the Faculty, the Vice-Chancellor's plan involved the loss of two academic staff, but following a Senate debate on his proposals, the Head of the Physics department, Prof. Chambers, offered an alternative plan which avoided closure of any section of the University, but which, as a consequence, increased the Faculty's loss from two to seven academic staff. Such a loss, in what was already a small Faculty in national terms, would have brought it dangerously close to the 'critical mass' which the Finniston Committee (Appendix 2) had considered to be necessary for the supply of the four-year MEng course to European standards. But in defeating the Chambers proposal, Senate would certainly have been forced to observe that the Faculty, together only with Law, had exceeded its Committee I target for savings. This had been set at £981,000, and the Faculty plan had exceeded it by £600,000, due largely to the ease with which engineering academic and technical staff could, and did, move to jobs in industry, carrying with them enhanced retirement benefits from the University through the UGC restructuring plan previously referred to. It was also pointed out, that even before these losses, the Faculty had consistently had a higher student-staff ratio than the Physical Sciences, coupled with a lower figure for the amount per student which the University had allocated to it (Chap. 8, Tables 8.4, 8.2 and 8.3). This disparity of funding between what were comparable parts of the University, had been known by senior members of the University for a long time from figures produced annually, but had become general knowledge as a result of the Vice-Chancellor's unit costs analysis.

---

[2] The author was a member of the UGC Technology Sub-Committee during 1980–9, and also Pro-Vice-Chancellor during 1981–4.

This disparity of funding was to surface again on many occasions; for instance, in February 1985 the Faculty board informed Senate

> That it is very concerned about the integrated effect over many years of University expenditure on equipment for Engineering which is very much less than for the Physical Sciences. For example, during the 3 years 1980-2 expenditure on equipment per full-time student was £960 in Engineering and £1570 in the Physical Sciences. This compares with national figures of £1132 for Engineering and £1334 for Physical Sciences.

As we shall see later, the Resource Allocation Mechanism (RAM), which was introduced in 1985, was a step towards financial transparency and more rational distribution of resource, but, as several Faculty deans discovered, it contained a sufficiency of mechanisms to allow it to perpetuate old habits.

This Faculty history is not the place to describe fully the turbulent, sometimes acrimonious, sequence of events which followed the final acceptance of the Vice-Chancellor's proposals, but it is the place to record the truly tremendous effort which he applied to the task of bringing the University successfully through these events. It took its toll, and he retired in 1984.

The history of the Department of Architecture is presented in the following chapter. The Faculty was responsible for creating it in the first place, played a significant part in its life in the University, and helped significantly to avoid redundancy amongst its staff when it was required to close in 1984. In Architecture, as in the rest of the University, there were no compulsory redundancies in any category of staff.

Chapter 7

The Faculty and the Department of Architecture, 1953–84

7.1 Introduction

This chapter has been included here because the Head of the Civil Engineering department from 1943 to 1968, Prof. Pugsley, played a pre-eminent role, significantly as Pro-Vice-Chancellor during 1961–4, in introducing architectural education into the University. His successor also played a large part in the attempts to integrate engineering and architectural education at university level, followed by a major role, also as Pro-Vice-Chancellor (1981–4), in the difficult period which ended in the closure of Architecture as a University Department. Many other members of the Faculty teaching staff spent much time during the 20-year period, 1963–83, in a sincere attempt to bridge the gap between the two disciplines, so that no history of the Faculty would be complete without a record of this failed experiment which it was responsible for initiating.

The experiment had been undertaken in an attempt to produce in the UK the type of architect – Prof. Luigi Nervi is usually regarded as the exemplar – who had emerged from several European Polytechnics, notably Turin and Milan in Italy, whose knowledge of structural engineering enabled them to escape from using conventional structural forms and ideas, allowing them to produce outstanding buildings, bridges and other structures. They had moreover, throughout Europe usurped the role of the structural engineer in many instances as the leader of the building construction team through their understanding of the principles of both structural engineering and architecture, obtained in a lengthy educational process which allowed students to take a first degree in both disciplines[1].

Readers of this history will recall that on several occasions the Faculty was tempted to take an interest in subjects related to architecture. In 1929, for example, the Dean's Annual Report contained the following statement:

> The next development which should be undertaken is the
> provision for students who wish to take up building science.
> The Department of Building Research emphasises the need for
> these facilities and the Engineering Board is of the opinion
> that the University could, with relatively small addition of staff
> and equipment, establish a satisfactory course.

The Dean at this time was Andrew Robertson and Sutton Pippard had just joined as Head of Civil Engineering. Both were members of the Government's Department of Building Research, which had a former student, R. E. Stradling (Chap. 4.2) as its Chairman. However, other developments took precedence and no action was taken.

7.2 The Board of Architectural Studies

The first meeting of this University Board took place in September 1953, but it must be supposed that even though it was attended by the Vice-Chancellor and chaired by the then Dean of Arts – the blind and unique Professor of Imperial and Commonwealth History – C. M. McInnes, the membership of Pugsley was significant. He was at that time a member of the Council of the Institution of Structural Engineers (ISE) during 1951–7, becoming its President in 1958. Then, as now, there were strong professional links at all levels between the ISE and the RIBA (Royal Institute of British Architects), involving discussions between them on the educational needs of intending practitioners in each discipline if they were to continue to collaborate professionally after the Second World War. Pugsley would certainly have played a part

---

[1] It is of interest to note that this practice was discontinued after the 1968 so-called 'student revolution', which in Milan and Turin was said to have been generated by architectural students. The engineering faculties at both polytechnics subsequently refused to accept students who had already taken a first degree in architecture.

in these discussions, and as subsequent events were to show, he held firm views on the need to include a greater input of engineering science into architectural education.

The Board decided to meet once each term, with business directed by the University Ordinance which made provision for association with other institutions so as '... To supplement the teaching of the University in such branches of professional or technical knowledge as they may deem fit.' As a first step, Mr. E. Freeth, Principal of the Royal West of England Academy School of Architecture (RWA) and a member of the Board, was accorded status as a Recognised Teacher of the University.

There was a second meeting at the end of that year, but not again until March 1955 when it was announced that the RWA had been assessed by the RIBA for the purpose of recognising their courses as satisfying its professional examinations, but had failed to do so; the grounds being that too few RWA students had passed these RIBA examinations for professional status in recent years.

7.3 The 1958 RIBA Oxford Conference

This failure by the RWA clearly caused the University to lose its immediate interest in architectural studies because the next meeting of the Board did not take place until November 1958 shortly after the RIBA's decision had been reversed, when it was called together to discuss, with the Dean of Arts in the chair, the report of the RIBA Conference on Architectural Education held at Magdalen College, Oxford in April 1958. The RIBA itself had publicly commented on the unsatisfactory nature of UK architectural education in 1956, and it is surprising that it took two years to organise a conference, the recommendations from which had such profound effects.

The conference report stressed that standards of competence should be improved at all levels, starting with standards of entry to courses of architectural education, which in many cases was no more than five passes at O-level. In fact, the first recommendation of the conference was that 'The present minimum standard of entry into training is far too low ... and should be raised to a minimum of two passes at 'A' level.' It is particularly relevant for this present history that a strong case was made for the development of new Schools of Architecture in universities and for the transfer to universities of Schools then existing in other institutions. Indeed, the third recommendation of the Oxford report was

> Ultimately, all schools capable of providing the high standard
> of training envisaged for the architect should be recognised by
> the RIBA and situated *in universities* (my italics), or
> institutions where courses of comparable standard can be
> conducted.

Not specifically written into the last part of this recommendation – but implied – was that architects should in future be educated in a high-quality, multi-discipline, academic environment. In making such a recommendation it was recognised that bridges between Architecture and other University Faculties would have to be established, particularly with Engineering, Science and Arts, and that universities would demand the development of architectural research activities as a condition of association with these Faculties.

It is interesting here to step aside from architecture for a moment, and to note that during the 1950s the two other major construction industry professional institutions, the Institution of Civil Engineers (Ref. 13) and the already mentioned Institution of Structural Engineers, had both reviewed their membership requirements, strengthening the intake standards for university courses which they were willing to accept as a component of their professional membership qualification. Also, the ICE, accepting that modern civil engineering required input from graduates in many disciplines, were now willing to accept degrees in physics, mathematics, biology, economics and so on, subject to an over-riding condition that a sufficient period of work experience in the civil engineering profession, or in relevant research, had been undertaken.

7.4 The University's Interest in Architectural Education

Returning now to the University's interest in architectural education, in December 1959 the Vice-Chancellor, Sir Philip Morris, convened a private meeting between himself, the Registrar (Herbert Butterfield), the Director of Education for Bristol (G. H. Sylvester), and the newly-appointed Director of the Bristol College of Advanced Technology (G. H. Moore[2]), which at that time had part-time day-release architectural students who worked in local practices. The Vice-Chancellor

[2] When the CAT became Bath University, Moore was its first Vice-Chancellor.

informed this group that the RIBA Oxford Conference Report had initiated a study of architectural education jointly by the University Grants Committee (UGC), the Ministry of Education and the RIBA itself. Sir Philip then made the surprising comment 'that the University had no ambition in relation to architectural education but was concerned merely to play any part which was necessary in the public interest.' Moore and Sylvester, on the other hand, no doubt wishing to strengthen the activities and importance of the fledgling CAT, hoped for a future role with full-time students.

In view of the above remarks by the Vice-Chancellor, it is perhaps surprising that an internal university group met as early as February 1960 with Pugsley as chairman and membership consisting of the Vice-Chancellor, two leading Professors from each of the Science and Engineering Faculties and one from the Arts Faculty. This group was quickly and formally recognised by Senate as the Committee on Architectural Education, being given the remit 'To consider the part which might be played by the University in the future of architectural education in Bristol.'

With Pugsley as chairman, and bearing in mind his detailed knowledge as a structural engineer of what the UK required in architectural education, and his subsequent pivotal role in this matter, it can be conjectured that he had persuaded the Vice-Chancellor to change his mind, or at least to soften his objections! Pugsley's enthusiasm for the task which he had undertaken is clear from the large number of meetings he held between February 1960 and January 1961, together with the detailed report which he presented to Senate on the latter date, a report which, in substance, was later sent to the UGC at its request, as the essentials of the University's intentions. The following is an abbreviated form of its recommendations.

(1) Steps should be taken to set up a School of Architecture in the University. It should draw upon established departments particularly in Arts, Science and Engineering.

(2) A three-year first degree should be provided, treating architecture as a broad academic discipline rather than a technology. For those seeking to enter the profession of architecture, this course would be followed by a two-year course acceptable to the professional registration body. Such a course should include both professional studies and research.

(3) A Chair of Architecture should be established in the University. A board of Architectural Studies should be established, to include representatives of the Faculties concerned together with professional architects.

(4) Admission to the School would be restricted to candidates qualified to take a University course and acceptable to the University.

This Pugsley report anticipates, and provides for, all the details of absorbing an existing organisation, the RWA School of Architecture, into the University, including such details as students' fees, salaries for staff members, arrangements for appointing new members and the retirement of some of the present staff. For a prescribed period, financial arrangements between the two bodies needed agreement, after which the University would be fully responsible. The accommodation of the School was also set out in the report, with the continuing acceptance of 25 Great George Street as the immediate solution. That this might have a not inconsiderable impact on future difficulties of complete integration of the school into the University, was not envisaged at that time.

Two immediate effects of this report were that the UGC asked the University to provide it with an estimate of costs, and that the 'Pugsley Committee' was enlarged by inclusion of representatives from the City of Bristol and the RWA, and thus became a 'Town and Gown' committee concerned with the implementation of the Pugsley Report. At its first meeting in March 1961, Pugsley reported that the UGC had no objection in principle to the University's proposal. Later that year the Committee on Architectural Education was revived and its visit to Great George Street was followed by discussions on whether it was possible that students taking the two first-degree system (see point 2 above) would be funded by Local Education Authorities and whether an A-level in mathematics, or a science subject, should be required for entry.

7.5 The Creation of the School of Architecture

As a first step in the implementation of the Pugsley plan the University set up a committee in May 1962 to appoint to what later became the RWA Chair in Architecture. At this juncture, Pugsley became the acting Vice-Chancellor due to the illness of Sir Philip Morris, and therefore took upon himself the task of specifying the particulars for the Chair, and selecting the assessors. In doing this, it is clear that he took sound-ings from his many professional friends about possible occupants of the Chair who would be supportive of his own ideas, and these resulted in an informal approach to Douglas Jones, Principal of the Birmingham School of Architecture. But Jones was content with that post and so declined the offer. The ensuing advertisement of the Chair had attracted 33 applicants by July 1962, but at the interviews of the five selected candidates[3] two weeks later, none were deemed satisfactory. As a consequence, Pugsley was given the personal task of finding a suitable candidate, and what inducements he offered to Douglas Jones we shall never know, but he did persuade him – apparently over lunch at the Athenaeum – to change his mind, and at the formal interview, which took place in November 1962, he was the only candidate. Although the Birmingham School was not one of the seven University Schools exist-ing at that time, it had full RIBA recognition, and as the future was to prove, Pugsley had found someone who would join him in creating his clearly expressed vision of architectural education, uniquely for the UK containing a substantial engineering-science component. Those who knew Pugsley, will here recognise the unusual strength of purpose and perseverance in such an apparently reticent man. He knew the type of architect he wanted to lead the new School, and was determined to appoint him, and no other.

Following the appointment of Douglas Jones, progress was quickly made and in February 1963 the Committee on Architectural Education was transformed into the Board of Architectural Studies with wide membership from the three faculties of Arts, Science and Engineering; of course with Pugsley as Chairman. One of its first decisions, which indicated the future involvement of the Faculty was the decision to require all new students to have obtained one of the many variations of the A-level in mathematics. That month also saw another step forward, in the agreement by the Ministry of Education for financial support of students taking the five-year split course, which was a little later agreed to be a three-year course for a BA in Architecture, followed by either a BArch or a BSc in Architecture, each of two years. The Registrar was given the task of arranging the financial aspects of integration with the RWA, with Pugsley and the Vice-Chancellor choosing the date at which

---

[3] With reference to later developments, it is noted that this list included a Mr Ivor Smith.

the RWA was to be integrated with the University, eventually deciding on November 1963.

For the next five years the insertion of engineering-science subjects into architectural education took place gradually, the specific contributions of the Faculty being largely from Civil Engineering and Engineering Mathematics. From the former, new courses in structures, foundation engineering, concrete technology and surveying were designed in collaboration with academic colleagues in Architecture. In a similar way, courses in mathematics were tailored to the existing knowledge of architectural students, and to their perceived future needs. There were significant contributions also from the Aeronautical and Mechanical Engineering Departments, both concerned with environmental aspects of building design.

7.6 The Onset of Difficulties

Pugsley retired from the university in July 1968, being succeeded by the present author. By November of that year the cumbersome and time-consuming structure of the affiliation of Architecture to the three Faculties of Engineering, Arts and Science had become clear, especially the need to submit examination results to each one for approval. A decision was therefore made to replace three by two, Social Sciences replacing the Arts and Science Faculties. It is to be noted here that it was not until the 1973 session that a majority group of the academic staff of Architecture proposed that their aims could be more satisfactorily realised if they were a Department of the Faculty of Engineering only. Their future problems, and eventual closure, might well have been avoided at this juncture had Douglas Jones, who unfortunately was due to retire at the end of the 1974 session, not expressed the view that it would be imprudent of him to commit his successor to this definitive change. Some differences of opinion, and possibly tension, amongst the academic staff on this issue are evident from the fact that a formal proposal was made at the same time to transfer the teaching of mathematics to architectural students from the Faculty to Architecture itself. This proposal, which was not accepted, was an indication of a serious issue which had emerged almost from the beginning of the teaching links between the Faculty and Architecture. It arose from the fact that, in order to indicate to architectural students that they must take the engineering topics as seriously as the overtly architectural topics, academic staff from Architecture were invited to join Engineering colleagues in helping students with problems in tutorial and discussion classes. For many, but certainly not all, architectural academic staff, this turned out to be an embarrassment, because their own knowledge of these subjects was usually inferior to that of their students, and, in consequence they made many different excuses for not attending the tutorial classes. One unfortunate effect of this failure to collaborate effectively, was the high failure rate amongst architectural students in these engineering and mathematical subjects in the summer examinations, requiring them to resit in the autumn, with consequential loss of practical experience opportunities during the Long Vacation.

During the latter part of the 1960s the UGC questioned the expensive nature of architectural education nationally, asking for justification of the need for five years of what was thought by many as a rather leisurely progress towards professional accreditation by the RIBA. As a result, the Bristol School, following the lead given by some others, reduced the total length of its courses for the 1973 session to four years, three for the BA degree followed by a diploma after one year of practical training for those who wished to become architects. For others, the MSc course was approved by the Science Research Council for an allocation of funded students.

The School also faced the implications of pressure by the UGC to consolidate architectural education nationally into a smaller number of larger Schools, and for its own survival asked the University in 1973 to allow it to expand, first to a total of 200 students and then to 300, with concomitant increases in academic and technical staff, coupled with a second Chair to be in the area of functional design. A principal argument behind this request was that Architecture was a discipline which covered the total built environment and needed not only to call on expertise in other parts of the University, but also needed a large complement of part-time specialised staff who were practising architects. This desire for growth was understood, but many in the University realised that the original 'Pugsley' concept of architectural education – having a science and engineering base, was fast disappearing. The proposed linking of Architecture to the newly created (1972) School for Advanced Urban Studies (Appendix 3), directed by Sir Colin Buchanan, gave further credence to this concern. As it happened, this concern did not materialise in any strength because SAUS, housed in Rodney Lodge in Clifton, had its own agenda which centred on public policy issues rather than the urban problems for which it had been created, and which were of concern to Architecture and Engineering.

7.7 The Beginning of the End

The appointment in 1975 of Ivor Smith to replace Douglas Jones did nothing to assuage the fears of those who wished to keep the Bristol School of Architecture in its role of providing a special, and nationally much needed form of architectural education. It is recalled that Ivor Smith had been one of the five interviewed candidates for the initial appointment to the Bristol Chair in 1963, none of whom had been considered as suitable for the more scientifically-based education which Pugsley was determined to implement. But Smith now persuaded the appointing committee that he was committed to continuing the programmes which Douglas Jones had begun according to the Pugsley specification, and so was appointed against two internal candidates who had been instrumental in developing and implementing this specification for the previous 13 years. Smith's approach to architectural education was well-known by its practitioners, however, and these two internal candidates quickly found chairs in other universities; Ben Farmer as Head of the Leicester School and Peter Burberry to a Chair of Building at UMIST. Both were senior lecturers who had been awarded MA degrees by the University in 1972 in recognition of their achievements.

At his first meeting with the Committee of Professors on Architecture, Smith gave his opinion that the School had concentrated too much on technical aspects, and on the strength of his claim that he was the only member of staff who had actually designed and built a real building, he argued that the School needed an infusion of practising architects. Clearly, he had not accepted that Bristol intended to be a unique School of Architecture with a reputation for education beyond that required for standard design of simple buildings. He also helped the School not at all by saying at the same meeting that '. . . architecture differs from other university disciplines in two ways – in the nature of the subject and in the nature of the course.' With regard to the latter, he expected other departments to provide what for them would have been low-level service courses for architectural students.

In 1976 Smith presented a new course structure, together with the removal of the entry requirement of an A-level in mathematics. The response to this from the Engineering Faculty resulted in him making known his wish to discontinue his responsibility to both the Faculties of Engineering and Social Sciences, and to opt for the latter. In justifying these views and actions, he strongly criticised the recommendations of the 1958 Oxford Conference on architectural education, which, it will be remembered, initiated the decision by the University to merge with the RWA School of Architecture – a decision which had the purpose of providing at least one school in the UK, where students would be educated within a framework of architectural and engineering science.

The appointment of Robert McLeod in April 1979 as Professor in Creative Design did nothing to halt the slide of the School away from its original concept. In fact, after initial requests by both Smith and McLeod for affiliation to the Faculty of Social Sciences, they produced the concept of an entirely new Faculty of Continuing Education, made up of Architecture, Extra Mural Studies, Social Work and Social Administration and the new School for Advanced Urban Studies. This suggestion coincided with a university-wide concern about the adventitious way in which Departments had become affiliated to Faculties as the University had developed, the result of which was the creation by Senate at the end of 1980 of a 'Working Party on Faculty Organisation'. Not surprisingly, Architecture made a very detailed submission, setting out all aspects of the various options which had been batted to-and-fro since it was created in 1963. The conclusion of the working party - that Architecture should be joined to Engineering in a new Faculty – 'The Faculty of Engineering and Architecture' – was not a surprise for those members of the academic staff in both the Faculty and the School who had continued to work closely together despite lack of enthusiasm from the School's management. This was particularly so with the research groups concerned with computer-aided design, with air-flow in and around buildings, and with energy use in buildings. The last of these research topics had grown to national significance through the appointment of Dr Brian Day, a senior lecturer in the School, as the coordinator of the Science Research Council's five-year programme on 'Energy in Buildings'.

It must be recorded, however, that the majority in the Faculty were strongly opposed to the formal linking of Engineering with Architecture in the title proposed for the new Faculty, arguing that it would give the totally false impression that, not only were they equal partners, but that the long-term emphasis which the Faculty had placed on engineering science was to be diminished in the future. But since Architecture itself was equally opposed to this new Faculty, the proposal died a natural death, even before financial issues made further discussion pointless.

7.8 The Effects of the 1981 UGC Financial Cuts

The two Professors of the School were clearly at odds with the majority of their staff; they not only entered detailed caveats about the principle of collaboration, but suggested a long period of assimilation before the proposed new Faculty came into being. However, it never happened, because in March 1981 the University was informed of the severe financial cuts which have been discussed in the previous chapter. Their effect on the School had consequences for the Faculty, and it may be considered as a historical accident that, as with the creation of the School in 1963, the Pro-Vice-Chancellor involved in helping to deal with the financial cuts was a Faculty Professor. He was also at that time a member of the UGC Technology Sub-Committee, which had the responsibility of advising the UGC Main Committee on how the 15 per cent cut should affect Engineering and Architecture at national level.

7.9 The Closure of the School of Architecture

From the beginning, the marriage between Architecture and Engineering had not been easy, but both partners persisted in attempting to overcome those differences of attitude to teaching and research which had caused European experiments in the dovetailing of the two university disciplines to be discontinued. Friction at academic interfaces is to be expected and is often productive, so that whilst Pugsley and Douglas Jones were in office a good start was made on integrating teaching activities, although there was only modest success in formulating common research ideas.

The School had always been housed in attractive accommodation in Great George Street, but it was many minutes walk from the Faculty, the academic staff common room and the students union. As a consequence of setting up their own social activities in Great George Street, there was reluctance on the part of both the academic staff and the students of the School to make the effort required to avoid isolation from the rest of the University. The effect of this failure to integrate satisfactorily, was that Architecture found itself vulnerable in March 1981 when the University faced the Government's cuts in both finance and student numbers.

Of concern to us here is the effect of these cuts on the relationship between the Faculty and the School. The UGC was not specific as to how universities should achieve the financial cuts imposed upon them, except with regard to student numbers. For Bristol the total reduction was 385, of which 315 were to be in 'arts' and 70 in 'science', the latter to include Engineering and Architecture. In addition, specifically for Architecture, the UGC had prescribed a national cut of 22 per cent on 1979 totals. The lead given by the Vice-Chancellor, Sir Alec Merrison, was that the necessary savings could only be achieved by academic staff cuts, following which it was accepted by Senate that the Faculty of Education should be cut from 32 staff to 18, and Architecture from 20 to 10. A part of the justification for this was that the costs of the Bristol School of Architecture were 30 per cent above the national average. Having made this decision, Senate passed its detailed implementation to the Committee of Deans who took the view that a School of Architecture having only ten academic staff was not sustainable, a view which leant heavily on the known UGC opinion that there were too many Schools of Architecture in the UK and that a viable size was 25 academic staff and 230–60 students. Another view expressed by the Committee of Deans is worth recording in full:

> . . . we are wholly unconvinced that the present Department
> has the unanimity of purpose in it which would be required to
> manage what we see as an enormously difficult transition. We
> therefore recommend to Senate the closure of the Department
> of Architecture, and again our recommendation is
> unanimous.

As an alternative to closure, the Deans recommended discussions of a merger with the Bath School. It was known that such a merger would be welcomed by the UGC, but Ivor Smith imposed the impossible requirement 'that the School would consider a merger with the Bath School but only on the basis of building up a larger Department in Bristol.' The likely reason for this antipathy between the two schools, is that from 1976 the Bath School had actually succeeded in developing an architectural science base to its teaching, given to it by Professor Ted Happold. He was an eminent structural engineer who broke away from the Ove Arup Partnership to found his own firm in Bath, in conjunction with the Chair of Architecture, specialising in the new field of lightweight structures. Not only had he been involved in the world famous, and structurally innovative, Sydney Opera House, but was later to have early responsibility for the Millennium Dome in London. His significant skill was in recruiting a team able to combine advanced structural engineering with superb architecture. Thus the Bath School had already gone even further than Bristol had achieved under the Pugsley/Douglas Jones partnership, and as previously explained, Ivor Smith had spent his period in Bristol in dismantling even this modest progress. It can now be seen as regrettable that the UGC did not bring serious pressure to bear on the two universities to force this merger, because it would have created, no doubt after some initial acrimony, a School of Architecture in Bath having international significance.

As a third alternative to closure or a merger with Bath, the Bristol School argued that it could viably exist with ten academic staff, but this was not accepted by either the Deans Committee or Senate. In February 1982 Council accepted a recommendation for closure of the School. As expected, there was opposition to this recommendation, not, in general, from academic staff in the University, who agreed with the Vice-Chancellor's vision that the only alternative was emasculation of several other departments in the UGC 'Science' group, but from some members of the University Court, who had the support of the Chancellor, Dorothy Hodgkin. At a special meeting of Court in April 1982 a resolution was passed by 92 vote to 90, requiring the appointment of a Special Committee of Court to review the decision to close Architecture. The members of this committee were the following:

Sir John Kendrew,	Chairman. President of St. John's College, Oxford, nominated by the Chancellor.
Sir Alec Merrison	Vice-Chancellor, ex-officio
Dr. Richard Hill	Chairman of Council, University of Bristol, ex-officio.
Prof. A. MacMillan	Prof. of Architecture, Univ. of Glasgow, nominated by the President of the RIBA.

Prof. A. M. Pritchard   Prof. of Property Law, Univ. of Nottingham, nominated by the president of the AUT.

Prof. R. T. Severn   Pro-Vice-Chancellor and Prof. of Civil Engineering, nominated by the Vice-Chancellor.

The Registrar (Evan Wright) and his deputy (Michael Parry), acted as secretaries.

This special committee met over the weekend 26–27 June 1982 and again on 29 July 1982. It visited the School of Architecture and invited presentations from Ivor Smith, who had already announced his intention to resign at the end of 1982[4], and from Michael Burton, the Head of Department designate. Because the committee had been asked by court 'to investigate the future of the department . . .', it interpreted this as an instruction to investigate the consequences of closure on the one hand, or continuation on the other. It did precisely that, which meant that the final decision was passed back to the University's chain of decision-making bodies.

During the discussions on the future of Architecture in the University, corresponding discussions on the future of Education had found a possible solution which envisaged a viable unit of the 18 academic staff originally proposed by the Vice-Chancellor, so that when Senate met for its autumn 1982 meeting to deal with the final decisions, any significant controversy centred on Architecture, specifically on the Report of the Special Committee of Court, presented to it by the Pro-Vice-Chancellor. Although there were continuing elements of opposition to the closure of Architecture, the majority of Senate recalled that during the past seven years it had – perhaps inadvertently – withdrawn from any real contact with other parts of the University, and had shown uncertainty within its own ranks with regard to Faculty affiliation. From these considerations, and possible alternative actions which would be damaging to the University as a whole, the proposal that it should be closed as from July 1984 met with no serious opposition. A further factor influencing this decision was, without doubt, the resignation of the School's two Professors, which not only indicated a lack

---

[4] The other professor, Robert McLeod, had taken voluntary redundancy in April 1982.

of cohesion and morale in the School, but presented the University with considerable immediate savings of possible redundancy or redeployment costs.

In dealing with affected staff, the Vice-Chancellor met each one individually with an AUT representative present to assure them that redeployment to other departments in the University would be the objective, but that for those older than 55 the existing early retirement scheme would be an option. In fact, the two academic staff attached to the computer-aided design group were transferred to the Geography Department, another, Brian Day, joined Mechanical Engineering and two technical staff, Terry Gorman and John Bracey, joined the same Department.

An interesting positive step was the move of Mr. M. Wells to the Department of Epidemiology and Community Medicine to create, with Prof. Colley, a Medical Building Design Unit to provide a one-year diploma course for overseas students. There were no compulsory redundancies, but this re-deployment of staff meant that only part of the expected saving of £300,000 p.a. due to the closing of Architecture was achieved.

The experience of merging engineering and architectural education was not entirely lost however, because the introduction of 'modularisation' of courses – to be described in the next chapter – allowed an optional 'unit' to be introduced into the second-year of the civil engineering course, having the intriguing title 'Engineering Architecture'. Its aim was to encourage the development of a professional interest in architecture as a closely related branch of engineering and environmental design. The syllabus and scheme of work of this unit, as described by its creator Ian Duncan – a structural engineer and part-time teacher in the Faculty, was to

> take a journey through the significant periods of architecture including Classical, Gothic, Renaissance, 19th Century and Contemporary. Students, working in pairs, will be required to research a subject of their choice in each of the areas covered. As well as presenting their discoveries, students will be expected to explore relevant background including technological, social, political and environmental.

Not surprisingly, this 'unit' proved popular in many parts of the

University, and in 2002 the Faculty was able to obtain the support of the Ove Arup Partnership in sponsoring an annual lecture (Appendix 5.6) on this theme.

Postscript

When university departments normally had only one Professor, he/she was automatically the Head of that Department for as long as they remained in post; they had the authority to initiate and direct the teaching and research programmes. This was the case in many Departments in Bristol in the early 1970s. Thus, in filing a vacant chair, the selection committee could decide that the new incumbent should be one who would take the Department to even greater heights along the same, or similar, intellectual track. Alternatively, for a Professor who had been less than successful, that someone with different motivation should be appointed. With hindsight, it can be seen that the committee charged with finding a replacement for Douglas Jones represented an unusually wide spectrum of academic interests, which either had no clear idea of the importance nationally of the experiment in architectural education which the University had embarked upon, or, thought that the experiment was misguided and wished Bristol to return to the conventional pattern of architectural education. Against national pressures to reorganise architectural education in the mid-1980s, there is no guarantee of course that an independent Bristol school would have survived, whether or not its engineering aspects had been maintained. A UGC-funded link with Bath would have been a probable outcome.

Chapter 8

Broader Horizons but Greater Government Control, 1984–90

8.1 Introduction

By 1983 the University had begun to adjust to its smaller size, although there was unease in what might have been construed as a loss in popularity amongst the schools, in that applications for entry had fallen from 23,853 in 1977 to 14,818 in 1983, but this was more likely to have been caused by the newer universities offering a range of less-demanding courses and attracting the less well-qualified applicants. From the Faculty viewpoint the most important event was the purchase by the University of the villas in the upper part of Woodland Road, which were to be used to re-locate the Arts Faculty from the Wills Memorial Building – making room for Geology to be located there so as to release the large space which it had occupied in Queen's Building since 1958.

In financial matters, the Committee I, referred to in Chapter 6, had a longer-term commitment, to make sure that financial targets were actually being achieved. It reported that in 1983 there would be a short-fall of £130,000, but by the following session an over-achievement of £300,000 was possible. However, there were some Departments which had failed to achieve targets, and a few which had over-achieved substantially. Amongst these few were the Departments in the Faculty. They had contributed appreciably to the reduction in academic staff in the University which had been a total of 165, in technical staff of 70, and in secretarial staff of 53, but the actual distribution of these losses amongst Departments was not conducive to the long-term academic health of the University as a whole. In the light of these figures, Committee I recommended posts for filling to a maximum of £100,000 for each of the sessions 1983 and 1984, which allowed the Faculty some relief from its self-imposed pressure. But Committee I also said it was disturbed to find that many of the under-achievers were still in that position despite having received an 'enquiring' letter from the Vice-Chancellor. In raising the possibility of enforced redundancy in these departments, it found it necessary also to recommend that over-achieving departments – such as the Faculty – should not be permitted to recruit to their full authorised staff levels until under-achievers had met their targets, either by retirements, resignations or re-deployment within the University. By 1985 the University had set up a Resource Allocation Mechanism (RAM) which listed various sources of income and expenditure in each department, and in doing so highlighted the under- and over-achievers. Table 8.1 is a short extract from the RAM data for 1986, indicating the large subsidy which the Faculty was giving to the Physical Science departments. Whereas most Faculty members regarded it as proper that they should give some financial support to the Arts departments, they considered that the Physical Sciences should do likewise. The financial position of the relatively small department of Geology was of particular concern at University level, and coupled with its modest research performance, caused the establishment of a special committee, with external assessors, to advise on its future. The outcome was phased early retirements and the appointment of new leaders.

8.2 New Blood Posts and the New Academic Appointments Scheme (NAAS)

The letter announcing the 11 per cent cut in resource which the UGC had been required to send in May 1981 contained a reference to £20m held back for 'restructuring'. Part of this process materialised in late 1982 as an offer of 'New Blood' posts, to be obtained on a competitive basis between all universities and all departments. In 1983 each post carried a recurrent grant of £21,000 for posts in natural sciences, medicine and technology – the UGC still using the last word to include engineering – and £16,000 for each post in the arts. An upper age limit of 35 was imposed, and in order to achieve its objectives, no existing academic staff were eligible.

The first opportunity to apply for these posts was in the 1983 session in which the University's list numbered 32, of which four were in Arts,

Table 8.1 University Resource Allocation Mechanism (RAM) for 1986. Expenditure and Income for Selection of Cost Centres.

£k	New Blood	Research Ctte.	Total Income	Total Salaries	Other Expend.	Income – Expend.
Chemistry	73	107	2299	1826	668	−195
Physics	73	63	1758	1526	314	−82
Geology			388	448	139	−199
Maths	49	17	886	800	46	40
Comp. Science			443	405	67	−28
Engineering	98	80	3081	2264	483	335

Notes
1. The UGC awarded 'New Blood' posts by competition.
2. The University Research Committee distributed funds according to RAE success.

24 in Science and Engineering and four in Information Technology. The Research Councils joined the UGC in assessing applications, and in the University 12 were successful, including one in Civil Engineering to expand the existing work in dynamic testing of prototype structures, and one in Aeronautical Engineering on aircraft systems and control engineering. The second tranche of these awards produced bids from almost every department of the University, but only seven were successful, one of which was for Mechanical Engineering in the subject of advanced manufacturing.

The New Academic Appointments Scheme (NAAS) followed the New Blood Scheme in a UGC announcement in May 1988. It was essentially of a bridging nature, providing for appointments in advance of future staff turnover, the assumption being that retirements would in due course meet the longer-term costs of the new appointments as the NAAS scheme was phased out over the period 1989–94. The total initial allocation over the five-year period was £70m, with a likely increase to cover salary awards during the last three years, and the University's share was £1.7m. Appointments were for new academic staff, significantly under 35 years of age, and normally at the lecturer grade, but could be higher in exceptional circumstances. As is indicated in Table 8.1, the Faculty had thrown a financial lifeline to the University during the difficult period in the 1980s by the voluntary loss of a large number of staff, and as a consequence of this, it was allowed six NAAS posts.

Two of these were allocated to the Communications Research Centre – one each in digital networks and in digital control; two to the Artificial Intelligence Research Centre – one each in parallel computation and logic programming; and one each to the Materials Research Centre and the Earthquake Engineering Research Centre.

8.3 The Report of the Steering Committee for Efficiency Studies in Universities (the Jarratt Report)

There were few Vice-Chancellors, or other senior administrators in UK universities, who did not realise in the early 1980s, that the rather loose organisational structures which had served them well before the expansions in size had taken place, were unsuited to the current and future situations. As a body, the UGC and its sub-committees also knew this from the visits which they made to each university at least once every five years, and more often when maladministration appeared to be occurring. Thus, when the Government began to insist on greater efficiency and transparency in the use of the modestly increased resources which it was now providing to the 46 publicly funded universities in the UK, Vice-Chancellors collectively saw no value in attempting to frustrate it – rather to be seen to be collaborating with it.

It was not surprising, therefore, that it was the Committee of Vice-Chancellors and Principals (CVCP), with Lord Flowers as its chairman, who appointed a very prestigious committee under the chairmanship of Sir Alex Jarratt, then chairman of Reed International plc. and Chancellor of Birmingham University, '. . . to promote and coordinate a series of efficiency studies of the management of universities and to report . . . on the results of the studies with comments and recommendations' (Ref. 14). In its membership of 12, there were four Vice-Chancellors, the Chairman of the UGC (Sir Peter Swinnerton-Dyer), one University Chancellor and one Master of a Cambridge college; the remainder were very senior figures from industry. It is evident, therefore, that any recommendations which the committee were to make, would be taken by the government to be generally expected by universities, and mostly acceptable to them. Indeed, it is likely that many Vice-Chancellors would have welcomed this help from above in directing their colleagues towards accepting change.

Six universities[1] were invited to submit themselves and their management procedures to expert scrutiny, resulting in a report in March 1985, originally directed to the UGC, but later to the Secretary of State for Education and Science, Sir Keith Joseph.

The report produced recommendations for government, the UGC, the CVCP, and for universities generally, with the greatest attention being placed on resource planning, allocation and management. One of its recommendations to Governments was that it '. . . should commission an examination of the structure and staffing of the UGC', and this resulted in 1988 in its replacement by the Universities Funding Council (UFC). In the University, a joint steering committee was set up to consider what action was necessary to deal with criticism that the Court, its supreme governing body, was much too big to be effective – it had 450 members – and that on its Council there were not enough lay members. It was clear that Jarratt was edging universities towards a more business style of management, with the Vice-Chancellor as a managing director supported by a small band of executive colleagues.

8.4 The University's Response to Jarratt

In the matter of resource allocation, the university moved quickly in the right direction by re-styling the Committee of Deans as a Planning and Resources Committee reporting to both Senate and Council. In response to Jarratt, the UGC had already become more transparent in its allocation of funds for both teaching and research, producing formulae which, in principle, the University was pleased to follow in its own distribution of finance to departments through the Resource Allocation Mechanism (RAM).

For the distribution of direct UGC funds to departments, the University used a 'unit of resource' for home and EC students which ranged from 5.65 for Clinical Veterinary Science to 1.00 for English and other Humanities, Social Sciences, Education, Law and Mathematics. The usual, unwarranted, distinction was made between the Physical Sciences at 2.10, and Engineering at 2.05. Clearly, historical costs, not current needs – or even performance – had been employed in calculating these figures. Using such units of resource, a formula of some complexity was employed to allocate to departments what in most cases

---

[1] Edinburgh, Essex, Loughborough, Nottingham, Sheffield and UCL.

would be their largest element of income. In it were components for the support of both teaching and research, the latter indicating the general obligation of academic staff to engage in this activity; it was therefore referred to as the staff research (SR) element of the direct UGC funding. For overseas students the process was much simpler, 80 per cent of the fee was given directly to the department.

Before leaving the allocation of teaching resource, it is noted that the UGC gave each university a quota of undergraduate and postgraduate students for each of its departments, and the allocation of resource depended on this number. The acronym MASN – Maximum Aggregate Student Number – came to be much used during later years. If a university allowed any department to exceed this quota, only the fee component would be paid for the excess. In 1987, for example, the Faculty's undergraduate intake quota was 205, but its intake was 247, and so 42 students were 'fees only'. Later that session the University asked the UGC for an additional quota of 800 students, some of which were specifically for the Faculty. The UGC refused this total request, but added, consistently with Government policy, that '. . . its Technology Sub-Committee is expected to consider sympathetically the bid for additional postgraduate numbers at its June meeting.'

For research funding, the University Research Committee allocated it according to three components[2], described as JR, DR and CR. The first of these – judgemental research –used money obtained by means of a levy on all departments based on student load and unit costs, subsequently distributing it according to its own judgement. For the first year, 1987, its judgement was to use the results of the 1986 Research Assessment Exercise, which meant that those 15 departments which had been judged as 'outstanding' received back 130 per cent of their contribution; the 16 'above average' 100 per cent; and average departments only 50 per cent of their contribution. On this basis it was not until the 1989 RAE that the Faculty JR income became positive. The DR – direct research – came from the overhead (40 per cent at the time) which the UGC paid to the University on the grant income from Research Councils and charities; one half of this was retained by the University and the other half given to the department. The third element, CR, refers to income from research contracts, which at that time

---

[2]   There was also a small JT fund, distributed by the Teaching Committee for innovative ideas in teaching.

carried a varying amount of overhead, but was later fixed at a minimum of 110 per cent of nett costs. For CR, the actual income to the individual department was at the discretion of the Research Committee, but was normally 80 per cent of the overhead element. Table 8.1 indicates that the Faculty received considerable amounts of both DR and CR funding. As time progressed there were changes to the details of these research funding arrangements, for instance, in 1990 the £60–70m DR part was transferred from the UGC to the Science Budget (i.e. the Research Councils). It was argued that this would be financially neutral because from this date Research Council grants would cover full costs, apart from academic staff salaries.

In both teaching and research the Faculty was a closely- knit unit for a substantial part of its activities, making it sometimes difficult to attribute a particular course or research grant to a specific department. Fairness dictated that the Faculty should have its own RAM, but this did not arrive until 1994. In the interim, a Planning and Resources Committee was created in 1989, having the responsibility for the allocation to Departments of resources given to the Faculty Cost Centre by the University; previously this task had been left to an informal meeting of the Faculty Professors. On other matters it gave delegated powers to three sub-committees dealing with teaching and undergraduate studies, research and postgraduate studies, and general purposes.

8.5 The University's Academic Plan

In addition to resource allocation and accountability, the Jarratt recommendations to universities was towards rectifying weaknesses in management styles and structures. As a means of assisting in this, in 1986 the UGC required each university to produce an Academic Plan, primarily for internal use in achieving the government and UGC ideas for rationalisation and change. Such plans were required to indicate proposed areas of expansion, consolidation, contraction or removal of activities. Of importance was that they should be consistent with known and expected financial resources. On this last issue the UGC was happy to report a very minor reversal of Government attitude to university funding; the block grant of £1212m already distributed for 1986 was to be increased by £14m, and that there would be increases of around 3 per cent in 1987 and 1988, with a similar further increase for the year after. These were, of course, inducements after many years of reduced awards, to those who were willing and able to satisfy the rationalisation programme of the UGC, but they were hardly sufficient to cover inflation, and did not alter the Government's declared aim of introducing greater efficiency by universities in using the resources it provided. Increases in student numbers were looked for without corresponding increases in funding.

The University's 1986 academic plan dealt principally with organisational and financial issues, which it became necessary to reconsider in 1989, with greater input from Faculties, and a forward look as far as the end of the century. There was no overt external pressure for this, although it was expected that the new Universities Funding Council (UFC) would want to review academic plans in 1990 after it had completed its takeover of the UGC. This forward look in the University envisaged a greater reliance on self-funding, from which emerged the 'Campaign for Resource', which will be discussed in detail in the next chapter.

8.6 The Faculty's Response to Jarratt – its Academic Plan

For the Faculty the fall-out from Jarratt was the establishment in November 1986 of a standing committee whose remit was 'To review the long-term structure of the Faculty including rationalisation, cross-departmental laboratories and all similar considerations.' Its chairman was R. W. (later Sir Robert) Wall, a Bristol Aeroplane Company (BAC) engineer, member of the University Council and chairman of its Audit Committee, and one-time leader of Bristol City Council.

Rationalisation was the key word used by the UGC; it meant either closure or merger of small departments – other than those few which were deemed to be vital to the national interest. In Engineering 'small' was defined for free-standing departments as those having less than ten academic staff, and for Unified or General Engineering departments less than 20–30. It was universally accepted that Bristol had a well-integrated Faculty, so that no individual department was in danger. This was fortunate because in 1986, unlike 1981, the Government told the UGC that it was not prepared to finance, even at 50 per cent, the cost of early retirements and redundancies. However, the UGC was itself prepared to help with rationalisation, because in the 1987–8 grant it held back £30m for '… redundancy and premature retirement to facilitate rationalisation and the recruitment of new, young academic staff'. These were the NAAS posts previously referred to.

A major issue for the Faculty's new Planning and Resources Committee was the creation of Faculty workshops and laboratories, rather than Departmental ones. This had been discussed many times during the previous 20 years but entrenched attitudes prevailed, now as before, with the over-riding difficulty that no money was available for effecting the change. As we shall see in Chapter 10, the Faculty had to wait until 2005 before an integrated workshop and laboratory system could be achieved.

In addressing the issue of student numbers, the Faculty noted that over the previous four-year period student numbers had increased by 30 per cent against a national decline, whilst maintaining entry standards at an average in excess of 26 A-level points. Subject to the provision of resources, particularly in accommodation, it expected to achieve a further substantial increase by the end of the century. Such an expansion would provide the economy of scale required by the UGC and would enable the Faculty to provide the breadth of subject matter required by the Professional Engineering Institutions. In detail, the planned expansion would be 20 per cent in Civil Engineering, 25 per cent in Aeronautical, 25 per cent in Electronics and 30 per cent in Mechanical. With these increases, coupled with four-year courses, the Faculty expected the undergraduate population to rise from 720 in 1989 to 1150 by 1995, and to 1600 by the year 2000. Tables A4.7 and A4.8 show that these were sensible targets, very nearly achieved.

In the forward look, on undergraduate courses there were to be more modular, interdisciplinary degree schemes and the provision of four-year courses. Building on the close Departmental links, these courses would bridge existing professionally-related disciplines. Subjects like avionics, space engineering, mechatronics and various combinations with electronics and computing would be attractive for both school-leavers, and for employers. To satisfy parity with European degrees, existing three-year courses would be extended and enhanced, incorporating management and business study options and a one-year period of study in continental Europe. Courses were to be modified as necessary to meet the needs of European students taking part in reciprocal schemes; this would be assisted if the University were to decide to adopt a 'credit-transfer' system for its courses and to move to a two-semester year – see section 8.11 of this chapter.

For postgraduate courses, a modular arrangement would be introduced in conjunction with industry, having the advantage of credit-transfer schemes which would be relevant to European student mobility. Its general view, though not without some dissent, was that full-time MSc courses were too demanding on academic staff time, although they were sometimes a source of research students.

In research, the Faculty had grants and contracts currently valued at £7m and looked forward to building on this figure if space and facilities could be made available. It had University Research Centres in earthquake engineering, information technology, materials, and advanced manufacturing and automation and two others were in prospect. In the plan, each Department listed its current research themes, together with those which it intended to develop; these are given in Chapters 11 to 16, which deal with specific departmental contributions to this Faculty history.

8.7 Development of Higher Education into the 1990s – a Government Green Paper

The above heading was the title of a government Green Paper published in May 1985 (Ref. 11), but preparatory to its publication in the latter part of 1983, it had asked the UGC for its views. In turn, the UGC had asked universities for their responses to a number of specific questions, some of which related to a further shift towards technological, scientific and engineering courses. When the Green Paper was published it asked how big a shift was possible in this direction, and over what timescale and cost. But with regard to future finance in general, whilst hoping to preserve 'level funding' for a time after 1984, the Government hoped that '. . . in the longer term universities should, in their own interest, be less dependent on public funds'. More specifically, it suggested to the UGC that it might consider the effects of '. . . a progressive reduction in the real level of such support in the order of, say, 5 per cent to 10 per cent per student overall by the end of the decade, and a further 5 per cent in the five years or so beyond that.' In the meantime the UGC was to consider a greater differentiation of funding between universities, and in doing so, to accept the possibility that some institutions of higher education might be closed or merged during the next ten years.

Taking the issue of funding differentiation first, the Faculty took the view that for the research activities of Departments, the dual-support system must continue. In this, the UGC contribution was at that time essentially in direct proportion to the number of students, whilst grants from the Research Councils were awarded on the merits of proposals made by individuals, or groups of researchers. Selectivity operated by the UGC should, the Faculty said, be based in the future on '... research ability to be carried out by a peer review system as is presently the case with Research Council funding'. The UGC was pleased to take note of this suggestion, which had been made by many universities, and in 1985 began preparation for the first Research Assessment Exercise (RAE), to take place in 1986, using its specialist sub-committees to place university departments in five categories in ascending order of merit from 1 to 5, which would determine the distribution of UGC funding for research. Further details of this RAE, and the five which followed it in irregular fashion until 2008, are given later in this chapter and in Appendix 6. Just as soon as the 1986 RAE was announced by the UGC, the University remembered that as long ago as 1982 the Advisory Board for the Research Councils (ABRC) and the UGC had, under the chairmanship of Sir Alec Merrison, suggested that each university should set up a Research Committee – or some equivalent which would satisfy the UGC – in order to allocate to its departments the research funds which the UGC provided. The University set up such a committee in June 1985, which included members of Council who had experience of industry and commerce, including research and development.

Concerning the increase in the number of students in science and engineering which the Government wished to see, and which they thought might be achieved by conversion courses at university for school-leavers who had taken arts subjects at school, the Faculty took the view that it would be a misuse of their highly-qualified staff for such courses, and that they already existed in Technical Colleges. Perhaps readers of this history will recall (Chap. 5.4) that in 1955, during a conference which the Faculty had organised on Education for Engineering, it had encouraged such 'arts' students to take the one-year intermediate course at Bristol for precisely this translation to engineering, but had discontinued it in 1966 due to lack of candidates.

A further point about the UGC/Government forward look into the 1990s relates to questions which it raised concerning the role to be played by the Engineering Council (Appendix 2) in accrediting engineering degree courses. The EngC was a child of Government intervention following the recommendation of the Finniston Committee (Ref. 4), and was being used by it in an attempt to subvert the role which had been played by the Professional Engineering Institutions and university-appointed external examiners for 150 years. The Faculty expressed its satisfaction with the existing situation and did not wish to see any other external body involved. Reference to Appendix 2 will indicate that the EngC did not succeed in reaching the overarching position which it aspired to, but transmuted itself into a registration body for the qualifications which the Professional Institutions continued to examine for, and to award.

Before leaving this forward look into the 1990s there are several further points worthy of mention. Prior to its actual publication, the government had, in November 1984, reduced student grants, withdrawing the minimum and requiring an increased parental contribution from higher income groups, including a contribution to tuition fees. The £39m saved was to be used to cover the continued growth in the student population, (£15m) modernisation of laboratories (£10m), and Research Council projects of high scientific and engineering promise (£14m). It had also announced financial support for a specific shift towards science and technology by earmarking £43m over three years for additional places in electronic engineering, and the Faculty was given additional students to benefit from this fund. A second point, is that not everyone in the University, and this included many in the Faculty, were content with the government's assertion that a significant part of the blame for Britain's poor economic performance lay with inadequacies in the higher education system, and its apparent assumption that only science, engineering and technology graduates could provide remedial stimulus. In a well-argued contribution, the Social Science Faculty produced many contrary arguments, which, in summary, said that what employers needed above all, was good graduates, irrespective of the subject studied.

Although the Government was looking ahead to the 1990s, its financial plans for the last part of the 1980s continued to be depressing for universities. Probably as a result of the Jarratt Report, from 1986–7 onwards they faced a squeeze on recurrent grant relative to costs of rather more than 2 per cent a year. In engineering and technology there was no doubt that the Government had listened to the needs of industry, commerce and the public services for highly trained and qualified manpower, and as a consequence the UGC had increased its 1989 undergraduate target in these areas by 1.7 per cent compared to 1984, but the government was not providing the means to achieve this. Here, the Engineering Council had certainly been performing a valuable service by repeatedly telling the UGC the historical fact of its underfunding of engineering relative to the physical sciences. In a letter to universities in May 1986 the UGC said:

> The average unit of resource for engineering cost centres has been *brought more into line* (my italics) with the average for the physical sciences. The Committee said in circular letter 14/84 that it shared the Engineering Council's view that engineering was underfunded. It noted particularly that the staff/ student ratio in engineering departments was significantly worse than the ratio obtaining in the physical sciences.

The phrase in italics in the above quotation indicates just how reluctant the UGC really was to properly fund engineering in universities. Its chairman, Sir Peter Swinnerton-Dyer, was a mathematician and, as pointed out earlier in this chapter, it continued to regard technology as synonomous with engineering and could not bring itself to accept parity between engineering and the physical sciences. Tables 8.2, 8.3 and 8.4 give numerical proof of the UGC's intentions in so far as it relates to the Faculty. Table 8.2 give figures for University expenditure on engineering and the physical sciences, for a sequence of years from 1971 when detailed accounts were first made available, in terms of unit costs per student for all activities except research. Table 8.3 tells the same story, but, because the University accounts were presented in a different format in 1990, fee income has been used to indicate student load, and the figures are for comparative purposes only, but give the same message as Table 8.2. Table 8.4 refers to the second point in the UGC letter concerning student-staff ratios, showing that it took until 2005 before reasonable parity was achieved.

But lack of UGC funding was not the only matter of concern for engineering education in the UK. Despite the allocation of earmarked

Table 8.2 Ratio of (Dept. Salaries + Laboratory Expenditure) divided by Student Load; comparison between Engineering and Physical Sciences

	'71	'75	'77	'79	'81	'87	'88	'89	'90
Eng.	0.94	1.61	1.80	2.34	2.95	3.50	3.74	4.21	4.21
Phy. Sci	0.98	1.78	2.30	3.06	3.71	5.72	6.14	6.84	6.26

Note
1. The unit is £k per student. Research income and expenditure has been ignored.

Table 8.3 Ratio of Expenditure divided by student fee income (ex research) for Engineering, Physics and Chemistry.

	1994	1997	2000	2004	2007
Engineering	n/a	0.91	0.96	0.95	1.09
Physics	1.42	1.16	1.18	1.28	1.24
Chemistry	1.29	1.43	1.58	1.55	1.70

Note

In 1990 the presentation of university accounts changed; fee income has therefore been used to indicate 'student load'. The figures given are for comparative purposes only, but give the same message as Table 8.2.

Table 8.4 Student/Staff Ratios for Engineering, Physics and Chemistry.

	1977	1983	1988	1992	1998	2003	2005	2007
Engineering	7.9	9.2	10.9	13.1	15.5	16.9	14.6	13.0
Physics	6.3	9.5	8.4	10.8	14.4	12.7	14.1	14.3
Chemistry	6.9	8.5	8.2	12.6	14.8	15.1	15.5	13.4

funds to information technology and for other specific areas of the engineering and technology programme, the number of graduates needed was substantially greater than the number of qualified candidates for places, and it was the opinion of the UGC Technology Sub-committee that this situation was unlikely to improve until the problems of early specialisation, and poorly qualified teaching staff in schools had been overcome.

Later in 1986 the financial situation became much worse due to rate and salary increases above those which had been budgeted for, causing the UGC to express an expectation 'that unless the Government provided more money, some universities would have to be closed'. In the same letter to universities as the one previously referred to, the UGC Chairman gave as his view that

> The Government is in no doubt that the quality of financial management varies greatly from one institution to another and that there were cases where it is totally inadequate. It would only take one catastrophe to make the issue of public concern.

Clearly, the Jarratt enquiries had made some unwelcome discoveries!

8.8 Extended And Enhanced Four-Year Courses

In March 1978 the Engineering Board reported to Senate

> In the light of the British Association Report on Education,
> Engineers and Manufacturing Industry, the Finniston
> Committee of Enquiry into the Engineering Profession, and
> the recent UGC initiative regarding four-year engineering
> courses, the Faculty intends to consider the long term nature
> of its courses.

No doubt the last phrase could have been expressed better, but the intention, given the context, is quite clear! There were at this time at least the three major reports listed in the above quotation, all pointing to what they saw as the deficiencies in the UK system of educating its top echelon of engineers. Although the Finniston Committee (Appendix 2) did not publish its report until 1979, its investigations into European approaches to engineering education were well-known, as was the fact that it would be recommending a greater distinction between the 'formation' – as it referred to the complete education and training experience – of the future leaders of the engineering profession, and those supporting them in less demanding roles. For the leaders, it proposed that resources should be provided by the UGC for the creation of four-year 'enhanced and extended' undergraduate courses leading to the Master of Engineering (MEng) degree; by enhanced it meant the inclusion of a greater content and range of topics, some of which the Faculty had always referred to as 'Engineer in Society' courses. With its experience of mobilising the high-level skills of university colleagues in other Faculties, coupled with the high quality of its student intake, it was natural for the Faculty to plan as the above quotation proposed.

But, as the previous chapter indicated, the financial gloom enveloping universities at this time was such as to inhibit new developments, and it was not until 1984 that Mechanical Engineering alone felt that it had sufficient resources with which to offer the MEng degree.

The four-year course, expected to begin in October 1985, was to be run in parallel with the three-year course, with the first two years having common material in the two courses, and selection for the four-year course was to be decided on merit at the end of the second year[3], the standard being set at such a level that around one-quarter of the 40–45 annual intake could expect to qualify. For any selected students who did not reach the standard set for the MEng at the end of the fourth year, the BEng degree could be awarded. An industrial advisory panel would assist the selection, and training in industry would be carried out during the summer vacations and would be funded by sponsoring companies. The initial selection of these companies leaned heavily on ICI and the nuclear power industry. Initially, the MEng degree was unclassified, but in 1991 a change was made so that it could be awarded 'with honours'.

At this time the Electrical Engineering Department would also have offered the four-year degree but for the fact of their commitment to starting the new three-year computer systems engineering course in 1984. Civil Engineering said that it would follow the wishes of its profession, which was the provision of further education for the brightest graduates after a period in industry, whilst Aeronautical Engineering, dominated by two very large companies, looked to them to provide additional education and training after its three-year course.

But the Engineering Council – set up as a consequence of the Finniston Report – had effectively determined that university engineering faculties would be divided into two classes; the smaller, more prestigious, class, offering a four-year MEng degree additionally to the three-year BEng degree, with the remainder offering only the BEng. It was clear, therefore, that the Departments of the Faculty must act together if they wanted to continue in membership of the more prestigious group. To do so required them to alter Regulations and Ordinances so as to change the title of the three-year degree from BSc(Eng) to BEng, and to introduce MEng for the four-year degree in all Departments, except for Engineering Mathematics which retained the BSc(Eng) for its three-year degree, and did not at first offer the MEng. These changes were accepted by Senate in February 1985, allowing the MEng in Mechanical Engineering to start in October of that year, but it was not until 1994 that the remaining Departments, including Engineering Mathematics, followed suit. By 1996 only Electrical Engineering had retained the BEng degree alongside the

---

[3] It had recently been decided that ten per cent of the second-year examination results would be counted in the classification for the BSc(Eng.) degree,

MEng, all other Departments concentrating their efforts on the MEng, which in addition to the four-year courses wholly in Bristol, now offered the two four-year courses described in the following chapter, which involved a year of study abroad.

8.9 Research Assessment Exercises (RAE)

The UGC method of allocating funding to universities, for both teaching and research, was based primarily on student numbers, which it determined for each subject area. After the 1963 Robbins Report (Chap. 5.6) the number of students grew, as did the number of universities, but the unit of resource provided to the UGC by the government fell progressively, and with it came the realisation that although the accepted duties of academic staff continued to be primarily teaching and research, not all of them were capable of making a valuable contribution to the second of these responsibilities. It should be said here that until the number of universities became too large for it to be feasible, the UGC main committee visited each one on a five-yearly basis, and its sub-committees did likewise to the relevant faculties. In doing so they were able to form personal judgements on both the qualities of the academic staff, and on the value of any research which was being carried out, but these were not allowed to affect the research funding distribution.

With government prompting, the UGC therefore began in 1985 to produce a mechanism for allocating its research funding according to the past performance and perceived potential of what it referred to as 'Units of Assessments' – in most cases an individual department – in UK universities. The first Research Assessment Exercise took place in 1986, being based on criteria which attempted to contrast 'input' against 'output'. Research grants awarded and university support in the form of laboratory and other facilities were the main input parameters, whilst output in engineering included the following:

1. published work, ranked according to the quality of the publication;
2. degrees, awards, medals and prizes indicating research quality;
3. evidence of national and/or international acclaim, such as editorial board members of prestigious journals, or chairman of a research committee;

4.   any indications from the professions on the value of the research.

As a result of long discussions between UGC members, and with assistance from the Engineering Institutions and the SERC, numerical values were attached to each of the criteria, allowing the 1 to 5 integer scale to be used in the final assessment, with 5 indicating some aspects of the research being at international level, and 3 being an average UK level. Actual details of this scale (and modifications to later RAE) are given in Appendix 6.

Much was learnt from this first attempt at research assessment and the UGC used it in its allocation of funds – referred to as the QR, or quality research element. As Table 8.5 indicates, three Departments in the Faculty were assessed as being above average, and two as average, which meant that the University benefited from the QR-funding innovation. Computer Science is included in Table 8.5 although it did not become part of the Faculty until after the 1989 RAE. Of course, there was some disquiet about the UGC's methods, but with necessary amendments, based on discussions with universities, the process was repeated in 1989 and 1992, with results for the Faculty as shown in Table 8.5. In the first two RAE, Engineering Mathematics was assessed as part of the University Department of Mathematics; in later RAE, Mathematics was itself split into Pure Mathematics and Applied

Table 8.5 Faculty Results for the first five Research Assessment Exercises.

	1986	1989	1992	1996	2001
Civil	5	5	5A	5A	5*C
Mech.	4	4	4A	4B	5A
Aero.	4	4	4A	4B	4B
E. and E	3	2	4A	5B	5A
Eng.Maths	3	3	4A	5B	5*A
Comp.Sci	2	3	4A	5A	5A

Note

In the first two RAEs the scale was 1 (low) to 5 (high). From the third onwards, Units of Assessment (normally departments) could, if they wished, submit only the 'research active'. On a descending scale, in percentages, A meant 95–100. B 80–94, C 60–79, D 40–59, E 20–39 and F less than 20. In 2001 the 5*grade was introduced to indicate that some sections of the submission were of exceptional merit.

Mathematics, with Engineering Mathematics part of the latter group; and until 2001, Mechanical and Aeronautical were considered as one unit of assessment. For the 1996 and 2001 RAE point 3 on the scale was divided into 3a and 3b, with a 5* added at the top.

Even before the 1992 RAE, some negative aspects of their effects had appeared, giving support to critics who pointed to the well-known principle that all methods of evaluation distort that which is being evaluated. Perhaps the most important of these distortions related to academic staff movements, from universities which had established prestige in research, to those newer universities which aspired to it, and to the QR funding which followed. Since each university could set its own professorial salary scale, there were many instances – more than one in the Faculty – where seriously good research groups lost their young and newly-appointed Readers, to Chairs in universities where they were expected to initiate research, not always with adequate facilities. Thus, in such cases, and from a national point-of-view, the effect of the RAE could be said to have been counter-productive, certainly in the short term. In taking part in such a process, a number of universities, having mortgaged their QR-future, were left with serious financial consequences when the actual assessments did not match their aspirations. In attempts to counter this poaching of researchers, there was an understandable reaction in the established universities, in the form of promotion to senior appointments, including Chairs, a good deal sooner than would otherwise have been the case, or achievements warranted. These universities, including Bristol, were also not averse to a little poaching themselves, and one finds in the Senate papers for November 1989 the following excerpt:

> . . . the University has to maintain a good 'share' of
> exceptional talent. This is not merely a matter of pride but
> increasingly a matter of survival. Such talent may not always
> be attracted through the normal system of recruitment or
> promotion. Moreover, the timing of such talent becoming
> available does not always coincide with the existence of
> declared vacancies.

A further effect of the RAE on academic staff appointments in Bristol was that at interview '. . . the contribution of the candidate to the Department's research profile' was to be taken into account, and if appointed, the later confirmation of the appointment was to be dependent on that contribution. It was also the case that staff in post who had '. . . an unsatisfactory publishing record were to be encouraged to improve their performance'.

Although the total volume of engineering research in UK universities increased following the introduction of the succession of RAEs, as evidenced by the number of papers published in an increasing number of journals, much of it was of an inconsequential nature, and there were many anecdotes concerning methods which some universities had used to obtain such publications.

For the first two RAE, the preparation of the necessary documentation was left to Departments, with some control in the Faculty by the Dean, but for the third, a University coordination was introduced through a panel of senior professors, all of whom had experience of UGC/UFC and research council organisations. The University's success in the 1992 RAE owed much to this panel and to its chairman, John Enderby (Head of the Physics Department), whose messianic enthusiasm ensured that each department made the best use of its achievements. The 1992 RAE was the first in which a choice could be made as to the proportion of academic staff to be submitted for assessment, and in Bristol a panel decision was taken that all departments should be in the A-category, which meant that the research of more than 95 per cent of staff was to be submitted. In the University as a whole the average score, from 45 submissions was 3.5, but in the Faculty (Table 8.5) it was slightly greater than 4. Table 8.6 indicates how the UFC used these results in allocating its QR-funding. For the 1992–6 sessions the quality multiplier was straightforward, as it had been since 1986, but for 1997–2000 there was greater selectivity, ratings between 3b and 5 attracting 50 per cent greater than the previous rating, with a 20 per cent premium for 5*. In the University, only Geography received this highest rating on the scale. For 2001-08 the selectivity increased still further, the lowest three categories receiving no QR-funding. One possible effect operating here was that of 'grade-inflation', which operated against concentration of research into the highest performing units. In Bristol, for example, in 1996 the percentage of staff in categories 5 and 5* was 31; in 2001 it was 51, with the Faculty percentages higher in each case! The opportunity to submit only a percentage of staff, coupled with the QR financial arrangements set out in Table 8.6, introduced a 'gambling' aspect into the process; because 5* was heavily rewarded – was it better to aim for this by submitting only the very best researchers, and if successful accept the reduced volume factor – or to include a larger number of researchers, receive a lower rating, but a better volume factor?

For the RAE which followed in 1996, 2001 and 2008, an 'Enderby' style coordinating group was created in the University, latterly with a Pro-Vice-Chancellor as chairman, but now more comprehensively supported by a section of the University administration referred to as RED – Research and Enterprise Development. Such an organisation was needed because the UFC, and the HEFCE which succeeded it, allowed the demands for yet more information to escalate to a level which brought into serious question the ability of the members of its subject assessment panels to read – let alone critically assess – all the material provided. Such was the concern generated by the Funding Councils, that in 2003 Sir Gareth Roberts – then president of Wolfson College, Oxford, and previously chairman of the CVCP – was commissioned to review the RAE procedures. However, he assumed that his task was to produce a more open and fairer system – rather than to reduce its complexity – which had the effect of increasing still further the amount of documentation to be produced. As an illustration of this, in the 2008 RAE, it was required for the first time that each researcher should submit copies of their four chosen publications for reading and assessment by the panel, and instead of each unit being given a single grade, it was to receive a 'research profile' for the academic staff submitted. Here, the research submitted by each member of the unit was to be placed in one of five categories, listed as 4*, 3*, 2*, 1* and unclassified, and presented in the form of the data shown for the Faculty in Table 8.5a, which is the actual result for the 2008 RAE.

On a scale of 0 to 4, the Faculty's average score was 2.90, giving it a position in the top five of UK engineering faculties. For the University as a whole, the corresponding score, for 48 units of assessment, was 2.72, which placed it fourteenth nationally in the HEFCE list. But such figures – described by the present author as 'notional rankings' – for both Faculty and University, cannot be considered as realistic because the HEFCE did not disclose, or use, the actual figures for the percentages of eligible staff submitted for evaluation by each university, and as noted previously for earlier RAE, some universities chose to make a financial gamble by offering for assessment only those whom they considered would be awarded the higher grades. The University chose not

Table 8.5a. Faculty Results for the 2008 Research Assessment Exercise.

	4*	3*	2*	1*	U.C	N.R.
Civil	25	55	15	5	0	8/23
Mech.	20	60	15	5	0	6/33
Aero.	25	55	15	5	0	3/33
E. and E.	10	55	30	5	0	17/35
Eng.Maths	25	45	30	0	0	3/46
Comp.Sci.	30	40	25	5	0	16/81

Notes
1. U.C. – unclassified; N.R. – notional ranking.
2. The first 5 columns give the percentage of those research staff actually submitted for assessment who were awarded the grade indicated in the first row.

to follow this option, but submitted 88 per cent of its eligible staff, the same as in 2001, but this actually meant a volume increase of 17 per cent. The real effects of the 2008 RAE will not be known until March 2009 when the HEFCE uses the results to distribute to universities the £1.2 billion of research funding. Although not officially stated, comments have been made indicating that one of the aims of this RAE was to use the ratings so as to allocate the greater part of the available research funds to about 15 universities.

During the 2008 RAE the HEFCE made new proposals for simplifying its assessment of research quality in science and engineering[4], aimed at removing the subjective nature of the peer review process by using 'metrics', such as the number of times published research was cited by others. No doubt it had been forgotten that in 1985 the UGC had considered this possibility in establishing the first RAE, but considered it to be not sufficiently reliable, because, amongst other disadvantages, it failed to recognise potential in younger researchers, and its objectivity was not always as secure as its supporters claimed. Much the same criticisms were levelled at the 'metrics' system in 2008, particularly by the Research Councils. It was argued by them that if the HEFCE could not evolve a more satisfactory RAE, then the existing dual-support system for research funding should be abandoned, with the Research Councils

---

[4] For other subjects it proposed a 'light-touch peer review', leading to its description as a 'twin-track' system of assessment.

Table 8.6 Mutipliers used by the HEFCE in allocating its research funding.

Grade	1	2	3b	3a	4	5	5*
1992–1996	0	1	2	2	3	4	n/a
1997–2000	0	0	1	1.5	2.25	3.375	4.05
2001–2008	0	0	0	0.31	1.0	1.9	2.82

taking over full responsibility for the total amount (£4.2bn in 2008) which government allocated for funding research in UK universities.

It is necessary to recall here, that in 2005 the Research Councils instituted a 'full economic costing' approach to grants, in which proposers were required to budget for additional items such as academic-staff time and costs of facility maintenance. For successful proposals 70 per cent of the total costs were awarded, leaving universities to find the remaining 30 per cent, most often from the QR funds for research which the HEFCE awarded. It was the declared intention of the Research Councils to raise their awards progressively to 100 per cent of total costs by 2010, which clearly would be possible if the HEFCE research element were added to their own. But the HEFCE were not to be dispossessed so easily from their part in the 'dual-support' system of research funding, and whilst the 2008 submissions were being assessed, it commissioned 21 universities (Bristol was not one of them) to carry out pilot studies to examine – using their own submissions –what differences might ensue between a 'metrics' approach and the existing 'peer-review' system. On this issue, the Faculty considered it to be important that universities should have an allocation of QR funds from the HEFCE, arguing that only they knew the young researchers in need of support, before they have a reasonable chance of obtaining Research Council awards. Amongst such awards there were now five-year EPSRC 'Career Acceleration Fellowships' for talented researchers who had three to ten years' postgraduate experience, the expectation being that they would have established an international standing by the end of the award. Similar fellowships were also available which released successful researchers from teaching and other duties, who 'had the potential to develop into the UK's international leaders of tomorrow'. In making such awards, which carried salary consequences, the EPSRC was already blurring the boundary between itself and the HEFCE.

8.10 The Science Research Park at Emersons Green

During the extreme financial difficulties of 1981, when the University was looking for longer-term sources of income, the idea of a science park was raised, with sites such as ICI land at Severnside being mentioned. In 1983–4, together with the City Council and Bristol Polytechnic, an attempt was made to establish such a park on land belonging jointly to the Council and commercial developers, but the initiative foundered, and it was not until 1987 that the universities of Bristol and Bath together with the Polytechnic formed a Science Research Foundation (SRF). It later became a limited company jointly with the Emersons Green Development Company, with the intention of developing a science research park at Emersons Green, an undeveloped area to the north-east of Bristol. The University's involvement was predicated principally on the involvement of the Engineering, Science and Medical faculties. For the Faculty, telecommunications, computing, high precision engineering, and new materials were listed as possible wealth-creating, high technological contributions.

The proposal was beset by planning difficulties, not only concerned with the science park concept itself, but also with the Bristol Ring Road development in that area – on which the science park depended for its viability, and with alternative proposals for use of the site. By 1997 it appeared that the science park would go ahead, but on the change of government in that year – from Conservative to the New Labour of Tony Blair, the new Secretary of State for the Environment, John Prescott, first 'called in' the planning consent, but then changed his mind and allowed the approval to stand.

In that same year the SRF produced a detailed business plan for an Academic Innovations Centre (AIC), which the universities of Bristol and Bath carried forward through a new company, Emersons Green Limited, and negotiated an agreement with the landowners for a 1500 m² building. By 2001 the Bristol Ring Road in that area had been completed, and in 2002 the Bristol City Region was awarded 'Science City' status by the Government, which prompted the South West Regional Development Agency (SWRDA) to acquire 12 hectares of land for the science park. In doing so they became a key player and in 2006 established a joint venture with two estate development companies. In the meantime, it was suggested that the imminent closure of the University's Long Ashton Research Station offered a possible site, but this was sold for housing development. Thus, the science park at Emersons Green has been under consideration for more than 20 years, during which time the accommodation occupied by the Faculty in the main precinct has been greatly enlarged, particularly for those departments and activities which were initially interested in a science park on the edge of the city. As is described in Chapter 9, the Departments of Electronic and Communications Engineering, and Computer Science, now occupy the greater part of the Merchant Venturers Building, and will move into the remainder as their research income justifies. Also, in 2005 all other research activities in the Faculty occupied new, extensive and high-quality space as a result of the BLADE project major reconstruction of Queen's Building (Chap. 10.3).

The possibility of a science park at Emersons Green had not been forgotten however, and in April 2008 the SWRDA was able to announce that the construction of a £300m 'SPark' would begin there later that year in collaboration with specialist developers Quantum Property Partnership and the universities of Bristol, Bath and the West of England. It was estimated that the 10-year project would eventually provide 6000 highly-skilled jobs.

8.11 Modularisation and Credit Transfers

In the Hale Committee Report on Teaching Methods (Ref. 9) and in the Government Green Paper (Ref. 11) discussed earlier in this chapter, there were suggestions of 'credit transfers' in university courses, by means of which students could move between departments, faculties and universities, carrying with them certification of having successfully taken and passed various courses. Universities were at liberty to accept or reject such 'credits' as part of their degree requirements. In the Faculty this had actually been possible since its first session in 1910, because the Ordinances and Regulations '. . . permitted the acceptance of studies elsewhere in lieu of periods of study at Bristol', so long as the final year and its examinations were taken in the Faculty. Also, within the University it had always been allowable for students to take courses in different faculties, but this was not common in engineering because the intensity of its degree programmes made it difficult to make the necessary arrangements. There was, of course, a very significant new dimension to this matter – the links with European universities described in the next chapter, and which, as we shall see, requires agreements on the equivalence of courses between the Faculty and the European partner.

In March 1989 the University set up a working party with a remit, falling within the Teaching Committee's responsibility, to review existing arrangements for modular courses and credit transfer schemes, and to make recommendations for the extension of these arrangements. It defined its task in the following terms;

> To develop, for introduction within a timescale of three years, a modular system into which all courses within the University can be fitted; the system to encompass credit transfer values, student load values and inputs into the resource allocation, all of which would be normalised across the University.

In setting its objectives, the working party, with Pro-Vice-Chancellor Alan Read as its chairman, noted that the 1989 University Plan envisaged a future in which it would be necessary to carry out effective teaching to larger numbers with fewer staff, and to assist in this, it considered that it would be valuable if a modular and credit transfer system could be devised for all University courses. To begin with, it was necessary to develop a common language, defining such terms as degree, programme, course and unit, the Faculty interpretation of which is indicated in Fig. 8.1, with 'modularisation' encompassing the total process. A major decision which the University made, on the recommendation of the working party, was the use of the Credit Accumulation and Transfer Scheme (CATS) which had been developed by the Council for National Academic Awards (CNAA) – the accreditation body for the Polytechnics – and which was being widely adopted throughout the UK. This scheme required that a numerical credit value be attached to each unit – which was given a unique University description, such as 'a204' for the subject Flight Mechanics in Fig. 8.1, and that across the whole University 120 credits would be the minimum for a year of full-time study. The really difficult task, that of actually doing this, was given to Faculties and Departments, and it needs little imagination to sense the uncertain nature of some of the numerical results.

Of all the Faculties in the University, Engineering was probably the most supportive of the modularisation scheme, essentially because its European initiatives had already indicated the need for some mechanism through which educational and professional experiences could be

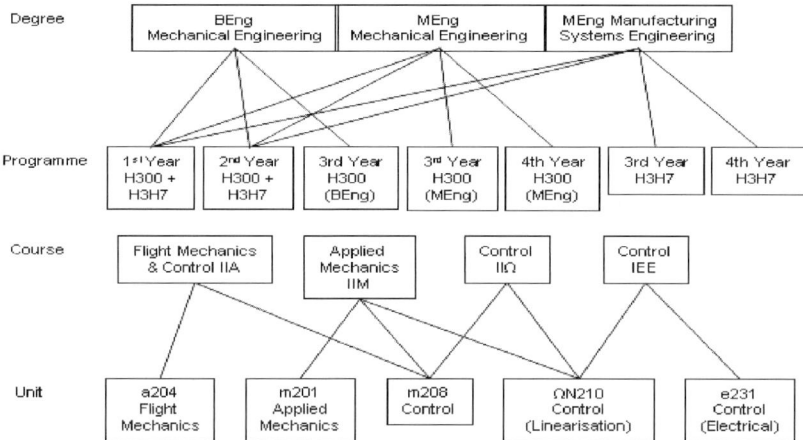

Fig. 8.1 One of the beneficial effects of modularisation – a single 'programme' being part of several degrees, and a single 'unit' being part of one or more courses.

measured in what was in the process of becoming a global economy. In its own work the Faculty was familiar with methods which attempted to quantify variables of a non-numeric nature, and so undertook the task of 'modularising' its teaching programme as a challenge. Even so, when it was required to make its report to Senate, it did so in the following words:

> This has proved to be a non-trivial exercise. Most of the
> Faculty's degree courses share common subjects, particularly
> in the first two years. Typically the students attend the same
> lectures but have different laboratory work, and hence
> technically they attend different units. We have several cases
> where we have four or more nearly identical units which will
> nevertheless appear separately in the course catalogue. A true
> description of the Faculty's teaching would have to be built up
> from elements rather than units.

For the benefit of others in the University, it went on to give an insight into its procedures:

> The original recommendation was that the credit for a unit
> should be proportional to the workload on a student doing
> that unit. We estimated the total workload for each first-year

unit, usually by a formula but with some subject-specific variations; and discovered that different degree courses had significantly different total hours, leading to different credit values for the same unit. Common credit values were obtained by arbitration and rounding off, from which it emerged that 120 credit corresponded approximately to 900 hours of total study time. This ratio was then used to assign credits to the remaining units.

It is not known whether the rest of the University followed this lead!

One spin-off from modularisation in the Faculty was that it did cause academic staff to liaise with colleagues on the detailed contents of their courses, and, where possible, to combine courses in basic engineering disciplines, such as applied and fluid mechanics, control, and theory of structures, where the growth of the Faculty had resulted in these subjects being treated separately by relevant Departments. The bottom line of Fig. 8.1 illustrates this point; the units in this course are given by four departments, the letters 'a' , 'm', 'e', and 'Ω' signifying Aeronautical, Mechanical and Electrical Engineering, and Engineering Mathematics, respectively.

Not all Faculties in the University were convinced of the value of modularisation, the Deans of Arts and Science both being of the opinion that it was academically unsound and would lead to lowering of standards, but their Faculty colleagues were divided on the issue. The Medical Faculty declared that its basic unit was a one-year course, which it could not divide further, so that it would give the 120 credits to each year of its five years of study. Apparently, all UK medical faculties adopted the same easy opt-out! This lack of unanimity in the University caused discussions on these issues to be prolonged beyond the three-year time-scale originally set, so that the decision to proceed was not made until May 1992, when Senate voted 46 to 16 in favour, with course structures throughout the University to be described in modular form in the Prospectus[5] for the 1994 session. An illustration of the Faculty doing its best to meet University decision-making is given in Table 8.7, which is for the three-year BEng degree in Aeronautical Engineering. Here, the total credit points for the first and second years are 121 and 119, respectively; for the third year, the total is either 121 or 119 depending on which option is taken.

---

[5] From 1990 a University Prospectus replaced those previously issued by individual faculties.

Table 8.7 BEng. Degree in Aeronautical Engineering. Course Credit Arrangements.

Year 1 Compulsory Courses		Year 3 Compulsory Courses	
AENG 101 Aeronautics	11	AENG 301 Aircraft Aerodynamics	8
EENG 101 Applied Electricity IAM	12	AENG 301 Aircraft Propulsion	5
CSCI 101 Computing IACEM	2	AENG 302 Aircraft Structures	11
AENG 102 Design IA	22	AENG 305 Design Project	22
EFAC 110 Laboratory Course IA	14	AENG 306 Flight Mech. & Control	11
EMAT 106 Mathematics IACEM	21	EFAC 103 Prof. Eng. Studies	16
AENG 103 Mech. of Fluids IACM	9	AENG Research Project	32
EFAC 101 Prof. Eng. Studies	2		
MENG 105 Props.of Materials IAM	5	**Course Work**	
CENG 104 Strength of Mats. IACM	9	AENG 201 Aeronautical Laboratory	14
AENG 104 Structures & Mechanics	5	AENG 203 Design IIA	6
MENG 107 Thermodynamics IAM	9	MENG 206 Materials & Mech. Lab.	1
Year 2 Compulsory Courses		**Optional Courses:** Students may take	
EENG 201 Applied Electricity IIAM	12	either	
MENG 202 Applied Thermodyn.		EMAT 302 Mathematics IIIAC	15
IIAM	10		
AENG 204 Flight Mechanics	11	or any two courses from the following	
MENG 203 Control II	5		
EMAT 212 Mathematics IIA	12A	ENG 303 Astronautics IIIA	8
AENG 205 Mech. of Fluids IIA	14	AENG 304 Computational AeroDyn.	8
EFAC 201 Prof. Eng. Studies	5	EFAC 310 French	8
MENG 208 Props. of Materials IIAM	5	EFAC 311 German	8
AENG 206 Theory of Structures IIA	14	EMAT 301 Mathematics IIIAC(half)	9
AENG 202 Cranfield Course	10	MENG 407 Strength of Materials IIIA	

The introduction of the course credit system in the 1994 session was a major administrative task for all sections of the University. To some extent it had an integrating effect because it made possible a selection by students, subject to permission from their tutors, of a range of courses differing from the conventional. In the Faculty, not only were students given a wider range of engineering options, but were also allowed a small number, usually two, of 'free credits', being courses anywhere in the University[6] if they could be fitted into the student's main engineering course programme, and were thought by the Faculty Board to be valuable to the student's overall education. Examples of such

---

[6] A university regulation was that all undergraduates must be offered 40 'free-choice credits' during their course; but only two being taken each session.

'credits' were two actually offered by the Faculty, having the titles, 'The Entrepreneurial Option' and 'Building Your New Business Venture'.

This wide range of options[7], the associated administrative work of keeping records of the choices made by students, and their performances in these choices through examinations and coursework, gave added urgency to the development of computer software within the context of the UGC/UFC initiative on Management and Administrative Computing (MAC), a package which involved a course structure database, transcripts for students, student-staff ratios and all aspects of the Resource Allocation Mechanism. The expertise in the Engineering Mathematics Department in software writing was put to good use here, in producing a system to deal with all educational aspects of each student's progress through the University, including choice of courses, examination performance, and final degree results. Without such software it would have been impossible to cope with the course/unit options made available to students, particularly those who chose to intercalate a year of study abroad into their four-year MEng degree. The in-house production of this software, written for both undergraduate and postgraduate courses, allowed it to be maintained and enhanced each year to incorporate changes in degree programmes for each student, their performance in examinations and rules for progression. In the absence of this software, the credit transfer system would not have increased teaching efficiency – rather the reverse – and the greatly increased administrative load would have diverted academic staff from their primary duties.

From the very beginning of the modularisation discussions, there was a background question of whether the university should, at the same time, drop its three-term structure and move to the semester system. It was quickly realised that a simultaneous introduction of both these major changes would result in chaos, and so the semester concept simmered slowly, until, in 1995, a proposal was made that the 30-week academic year should be split into two equal parts, each having 12 weeks dedicated to teaching followed by three weeks for all other activities, which would include examinations. But it was not until a year later that a final decision was taken, that the Bristol form of 'semesterisation' would be two 12-week periods – described as 'teaching blocks' – with a final six weeks for revision and examinations. Notwithstanding this

---

[7]   Uncoordinated departmental developments had resulted by 2007 in the Faculty delivering 261 units in 52 undergraduate programmes. Modularisation had not solved this problem.

decision, the academic year continued to be split into three terms, with long breaks at Christmas and Easter!

Alongside these discussions on semesters, the opportunity was taken to use the experience gained on modularisation to relax the credit points system from 10/20/40 to 10/20/30/40. It had also become clear that some Departments had accepted modularisation more in the breach than the observance, by severely restricting the number of 'free credits' which they were allowing, and one Department was even allowed to have a final-year of eight subjects, each of which carried 15 credit points.

When the working of the modularisation programme came up for its periodic review in 2003, a suggestion was made that the University could 'brand' itself by following Princeton's requirement that all science and engineering students be required to take at least one arts or social sciences option, and that students in these last two Faculties must take at least one science or engineering option. The Faculty was able to point out that from its earliest days study of a European language had been required, and, more recently, the MEng involving a year abroad more than fulfilled the 'Princeton' objective.

8.12 Computer Science Joins the Faculty

The Senate minute of 30 January 1989 formally recognised that Computer Science, then a sub-department of Mathematics, had increasing links, both in teaching and research, with the Faculty departments of Electrical and Electronic Engineering and Engineering Mathematics. The minute recommends

> ... formal steps be taken to transfer responsibility for the
> Computer Science and Computer Science with Mathematics
> courses from the Faculty of Science to the Faculty of
> Engineering; transfer to take place for the 1989 session.

Some impetus for this change can be found in the action of the British Computer Society (founded in 1957) in seeking affiliation to the Engineering Council (Appendix 2). It was already able to award the title of Chartered Scientist (CSci.) to graduates in computer science, but membership of the EngC would give it the ability to grant them the more widely understood IEng and CEng professional titles. A further push came from the Alvey Committee, which had reported to Government on the needs for education and research in information technology. In 1983 it had funded a BSc degree in computer science, which in the Faculty had resulted in 1987 in a proposal that an MSc in 'Foundations of Artificial Intelligence' should be offered, this being a major area of research in Engineering Mathematics under the direction of Prof. Jim Baldwin.

The formal transfer to the Faculty took place in 1989, but Computer Science had split from Mathematics in 1984, becoming a separate department, and although it was still part of the Science Faculty, in 1985 it transferred its accommodation to Queen's Building, occupying space previously occupied by Geology. Its first Head was Prof. Michael H. Rogers who, since he joined the University in 1958, had been a key figure in promoting computing as a service to the rest of the University, becoming Director of the Computer Centre in 1975. But as its activities became increasingly administrative, he returned to Mathematics as its Head of Department, coupling that with directing the Computer Science Group, and it was in this role that he joined the Faculty.

At its beginning, digital computation was not recognised in the University as being of major importance, and the Faculty bore its share of the blame. In the early 1960s when a computer was allocated to the SW universities as a group, one of its senior professors[8] expressed the view that he could not see the computer having any part to play in engineering research, as a result of which the machine which was on offer went to Southampton University!

As younger professors came to have influence in the Faculty, the situation changed, notably in 1968 when Prof. Collar, then a PVC, agreed to become chairman of the University Computer Committee. However, the change went only so far as to develop activities in computing as a service, not as a science in its own right. Computing as a science was an activity persued by only some of the academic staff in the Computer Science Department in 1985, a fact which became clear in the result of the 1986 RAE, when Grade 2 was its achievement, and which was only raised to Grade 3 in the 1989 RAE which followed.

For the 1992 RAE the Department had grown to 13 academic staff, and since 1988 it had been awarded in excess of £2m in research grants, notably from the EU ESPRIT programme, but also from the SERC and

---

[8]   At this time, professors had research assistants to perform any calculations which were necessary and feasible.

industry; and in collaboration with Electrical Engineering and Engineering Mathematics it had created the Advanced Computing Research Centre. As a result it was able to report its success in three overlapping areas – logic programming, parallel computing, and machine intelligence and computer vision – which earned it the award of Grade 4A – the 'A' of course indicating that all academic staff had been declared as 'research active'.

In its undergraduate programme, in 1993 the Department joined the rest of the Faculty in offering a four-year course, in which an option was for the third-year to be study at a university in continental Europe, initially styled as MSci, but soon conforming to the general title of MEng, which later allowed the third-year to be taken anywhere in the world. The BSc degree continued to be offered, either in Computer Science, or in Computer Science with Mathematics, and in addition there was a joint honours degree in Computer Science and a Modern Language which was still offered in 1998, but discontinued soon after, leaving the others to be available right up to the 2009 session. One consequence of this demanding undergraduate programme was that in the 1994 HEFCE Teaching Quality Assessment (TQA) – for which the gradings were 'excellent', 'satisfactory' or 'unsatisfactory' – only the middle grade was obtained, and although the system permitted appeals, none was requested. Indeed, the Department seemed happy to accept a heavy teaching load, because in 1995 it offered a taught MSc course for the first time, the subject being 'Advanced Computing – Global Computing and Multimedia', and this was added to progressively until, for the 2008 session, a total of six taught MSc courses were on offer, although several had common components, and one was concerned with elements of computer science 'for graduates in other disciplines'.

The appointment of David May from INMOS in 1995, coupled with the move to greatly enhanced facilities in the Merchant Venturers Building, opened new horizons for the Department, allowing the 1996 RAE submission to contain topics and achievements previously not possible. With May and Warren as directors, it had more than £2m of research funding, including a £350,000 award from the Mitsubishi Research Institute for 'Developments in Integrated Parallel Processing Technology'. It had also set up a joint research centre with SGS-Thomson Microelectronics (formerly INMOS), and had taken under its wing the Faculty's Safety Systems Research Centre, which had recently received £500,000 of sponsorship from Nuclear Electric, the

Civil Aviation Authority, Railtrack and Lloyd's Register. It was now able to record that its research was organised into the following three themes:

> Computer architecture, including microprocessor and interconnect architecture, and parallel computer architecture;

> Declaritive systems, including the design and implementation of functional and logic programming languages, and programme analysis and transformation;

> Multimedia, including graphics, vision, speech and image processing and communication.

There was little surprise therefore, that a Grade 5A was awarded in the 1996 RAE, and with continuing growth in academic staff and research output, the same result was obtained in the 2001 RAE.

In its submission for the 2008 RAE the Department was able to justify its claim to be an international centre of excellence in the theory and application of computing. It pointed to its own involvement in the major changes which computing had effected in a very wide range of human activities, through its specific programmes in machine learning, graphics, cryptography and quantum computation. In this last area for example, it was able to point to its involvement in the University's new Centre for Nanotechnology and Quantum Information, which was being built with £10m of SRIF2 funding. This significant extension of its research activities had been made possible by an increase in academic staff from 29 to 37, indicating confidence by the University in its ability to attract major research funding, as, for example, the £12.6m joint venture with Electrical Engineering in the University Innovation Centre which links communication with computing and content – the '3C Research Centre' – and the spin-out Company XMOS which had secured £16m of venture-capital funding.

Chapter 9

Growth and Division – 1989–96

9.1 European and International Exchange Programmes

By the second half of the 1980s the Faculty had established links with many European educational establishments, notably in France with the various *Ecole National* corresponding to its own departments, and with institutions of similar status in Italy (Milan), Germany (Hanover) and Spain (Madrid). The introduction in 1987 of the TEMPUS and ERASMUS[1] programmes by the European Commission allowed these ad hoc arrangements to be both formalised and funded, and in the Faculty resulted in the creation of four-year MEng degrees in which the third-year was to be spent in an agreed European academic institution, with the student taking courses and examinations in that institution, and in the language of that country. These degrees were initially described as 'MEng with European Studies', which was soon realised as not being a sufficiently clear description that what was involved was a one-year period of study in a European university, and not a course of European Studies in Bristol! The name was therefore quickly changed to 'MEng with Study in Continental Europe' and for brevity it will henceforth be referred to as the EuroMEng. At first, only Civil and Electrical Engineering offered this degree, but they were soon joined by all other Departments in the Faculty. The ERASMUS programme was a reciprocal arrangement, requiring that an approximately equal number of

---

[1] TEMPUS – Trans European Mobility Programme for University Students. ERASMUS – European Community Action Scheme for the Mobility of University Students.

European students took appropriate courses in the Faculty, and this influx, taking lectures, coursework and laboratory classes, was of great cultural benefit to all students studying in Bristol, not only those in the Faculty. But these exchanges did not take place without a considerable administrative load on academic staff, notably on Dr Roger Moses, who in 1981 had been appointed the Faculty's Director of Continuing Education.

The opportunity to take part in the EuroMEng was restricted to students who had performed well in their first two years in the Faculty; this was originally set at a minimum of 55 per cent average examination mark, but was raised to 60 per cent after a few years of experience had indicated the academic level and personal qualities required. The student was also required to have obtained at least an O-level in the relevant European language, and be willing to take supervised tuition and an assessment of competence at the University Language Laboratory. In view of the demands which these exchanges made on the students involved, careful selection was necessary, but in almost all cases the outcome justified the initial choice, and on graduation with the EuroMEng many Bristol students became valued employees in the European country where they had studied, not only because they were potentially very good engineers, but also due to their ability to be at ease in both English and the European language.

Reference back to Section 8 of the preceding chapter will show that the four-year 'enhanced' MEng degree did not appear in the Faculty, except for Mechanical Engineering, until 1994. This simplified arrangements for the EuroMEng until that time, because the returning student, after a year in Europe, took the third-year courses of the BEng degree, for which the first two years had been a preparation. The only additional action required by the Faculty was to make sure that the academic and cultural experiences in the European institution were of a sufficiently high standard to merit the award of the more prestigious degree of the EuroMEng. In fact, almost all the returning students showed, by their performance in their final examinations, that they had made excellent use of their year in Europe.

After 1994, with the four-year 'enhanced' MEng degree in place throughout the Faculty, and the three-year BEng being phased out in most departments, the management of the EuroMEng became more complex, and therefore more demanding on the academic staff involved. The principal reason for this was the introduction in 1994 in the whole University, of the 'modularisation' of all its courses – a scheme which has been described in detail in the previous chapter. In essence, it entailed an assessment, ultimately in numerical terms, of the content and value of each course – or 'unit' in the terminology used. The data thus produced was openly available, enabling students to know, with precision, the contents of the courses they were required to take throughout the four-year MEng. For those students opting to spend a year at a European institution, the critical information for them was details of the third-year courses which they would be missing in Bristol, and whether the European partner could provide the same, or similar, material. If not, how was it possible for them to cope with the fourth-year Faculty courses when they returned?

Being totally convinced of the value of the EuroMEng, the Faculty solved this difficulty, with the agreement of the Professional Engineering Institutions as qualifying agents for the MEng on behalf of the Engineering Council, by introducing a course structure, the essential features of which was that the fourth-year courses now dealt with topics which did not depend directly on earlier success in the third-year courses. Table 9.1, which is for the EuroMEng in Civil Engineering, exemplifies such a course structure. The key statement here is that appearing below 'ListCENG3', which states that all five of the key subjects listed there must have been taken (and examinations passed) by the end of the fourth year. What it does not say, is that students were allowed *to decide for themselves* (my italics) whether their European studies had covered the greater part of these key courses. If they so decided, the Faculty accepted their decision; otherwise, these courses had to form part of their fourth-year studies, with corresponding reduction in the number of courses taken from 'ListCENG4'. Table 9.2 explains the system: taking the second name, James Craddock, by way of illustration, he had assured the Faculty[2] that his studies at INSA Toulouse had covered at least 60 per cent of the material in the two Faculty courses/units Structural Eng 3 and Geotechnics 3 and had passed the necessary examinations. In his fourth-year he was therefore required to take the three remaining units in ListCENG3, and because he had not taken a 'research project' in Toulouse, he was required to do this also in his final year at Bristol.

The note below 'List CENG4' indicates the point made earlier, that subject to formal approval by the Department, students were allowed

---

[2] In setting up the exchange, a Faculty representative would have visited Toulouse and discussed the necessary details of its courses.

Table 9.1 MEng in Civil Engineering with Study in Continental Europe (H201), or with Study Abroad (H202). 4th Year Programme.

Unit Name	Director	Unit Code	Credits
Civil Engineering Systems 4	JPD	CENG M1800	10
Structural Engineering 4	WMS	CENG 2400	10
Geotechnics 4	DFTN	CENG M2300	10
Hydraulics 4	JHL	CENG M2100	10
2 units from list CENG 3 (see below)			20
1 unit from list CENG 4 (see below)			10
Water Resources Project 3	MARR	CENG 31600	10
Major Project 4 (Research or Design)	JA	CENG M5100	<u>40</u>
		Total	<u>120</u>
List CENG 3			
Structural Engineering 3	WMS	CENG 31400	10
Geotechnics 3	MLL	CENG 32200	10
Water Engineering 3	SEH	CENG 33600	10
Civil Engineering Systems 3	SHE	CENG 33900	10
Professional Studies B	MW	EFAC 30002	10

If any of the above units has not been studied in the year abroad to a level acceptable to the Faculty, they must be taken instead of the units in List CENG4

List CENG 4			
Scheme Design 3	EI	CENG 36400	10
Seismic Analysis 3	NAA	CENG 32400	10
Timber Engineering 3	IHGD	CENG 32700	10
Earthquake Engineering 4	NAA	CENG M1900	10
Flood Risk Management 4	DH	CENG M3000	10
Geotechnical Modelling 4	DMW	CENG M1700	10
Power Generation for 22nd Century 4	GLQ	MENG M0002	10

Alternative Units may be taken with the agreement of the department.

Weighting for purpose of degree classification:	4th Year	70%
	3rd Year	20%
	2nd Year	10%

to take non-engineering courses – 'free credits' they were called – from other faculties in the University. Thus, the word 'enhancement' relating to four-year courses, was not necessarily in depth, but could be in breadth, either in an engineering, or more general, sense.

The financial provisions for the EuroMEng were attractive to students from the beginning, and later doubly so, when tuition fees in the UK reached high levels. All fees are paid by the ERASMUS fund, and are claimed from the European Commission by the two institutions

Table 9.2 An illustration of the course structure for students spending a year abroad.

		Units Done Abroad							Units to do this Year						
		Structural Eng 3	*Geotechnics 3*	*Water Eng. 3*	*Civ. Eng. Sys 3*	*PSB*	*Research Project*	*Design Project*	*Structural Eng 3*	*Geotechnics 3*	*Water Eng. 3*	*Civ. Eng. Sys 3*	*PSB*	*Research Project*	*Design Project*
INSA Lyon	Graham Sutton	x	x	x							x	x	x		
INSA Toulouse	James Craddock	x	x		x		x				x	x		x	
Valencia	Ben Pryke			x	x	x			x	x	x				x
	Jonathan Shepherd			x	x	x			x	x	x				x
Barcelona	Ian Baggs			x	x	x			x	x	x				x
Innsbruck	Thomas Ashworth	x	x	x	x		x						x		x
	Gillian Evans	x	x	x			x					x	x		x
	Sarah Harding	x	x	x			x					x	x		x
Hanover	John Hurle	x			x	x			x	x	x				x
Chalmers	Jay Edwards			x	x		x		x	x	x		x		
	Robert Jenkins			x	x		x		x	x	x		x		
Berkeley, Calif.	Andrew Machen		x	x	x	x	x		x						x
Irvine, California	Joe Walton	x	x		x	x			x		x				x
	Rupert Inman	x	x		x	x			x		x				x
San Diego, Calif.	Nick Bass	x	x				x				x	x	x		x
Univ. of West. Aust.	Stuart Rogers			x	x	x			x		x			x	
Toronto	Mark Chambers	x	x	x	x	x						x			x

involved. In addition, the student is paid a maintenance grant directly, which for a Bristol student in 2007 was £2800, together with travel costs for one return journey. The ERASMUS programme does not provide for non-European students studying in EU countries.

In 2005 Engineering Mathematics widened the opportunity available to students of intercalating a year of study outside Bristol during the MEng, by allowing it to be taken anywhere in the world, subject, of course, to appropriate arrangements, and in 2006 the rest of the Faculty followed suit. For what is now described as the 'MEng with Study Abroad' – for which the arrangements such as those set out in Table 9.1 also apply – students favour the English-speaking countries, particularly the USA, Canada and Australia, but, unlike the ERASMUS scheme, there is no organisation providing financial support, other than modest, and competitive, University funds to cover travel costs.

Table 9.3 presents Faculty statistics for both exchange programmes during the sessions 2003–07, from which it is clear that the ERASMUS programme brings more students to Bristol than the Faculty sends to Europe; one likely reason for this is the disadvantage to UK students of English being the international language, a facility in which is a common attribute of European students at an early age. The numbers in Table 9.3 may not appear excessive, but each of these students generates an administrative load greatly in excess of the norm, and as a consequence it has been found necessary to limit the incoming numbers. It is fortunate that the University International Office (UIO) is available to help departments in dealing with non-academic matters relating to these students. One of the tasks which the UIO imposes on ERASMUS students is that of writing a report on their experiences. These normally give a positive picture, but often indicate the need to manage, and adjust to, a totally different relationship which exists in European universities between academic staff and their students. The following extract is from the report of a Faculty student who spent a year at Aachen in Germany:

Table 9.3. ERASMUS and Study Abroad Students, 2003-07. In each main column outgoing student numbers are to the left, incoming students to the right.

	Aero.		Civil		Mech.		Elect.		C.Sci		E. Math.		TOTAL	
Italy		6	2	19		5		5		5			2	40
France	10	26	5	28	5	23	6	17	3	16	1		30	111
Austria		1	9	9		5				4			9	19
Spain		10	7	19	5	7	1	18		10		4	13	68
Czech.						2				10	1		3	10
Germany	3	8	2	5	3	6	4	12	1	4			13	35
Sweden	4	6	4		6	6							14	12
Portugal										12				12
Hungary												11		11
TOTAL	17	57	29	80	19	52	13	52	4	61	2	15	84	318
USA	9		12		13		2		12				48	
Canada	1		5		7								13	
Australia	3		8		10				1				22	
TOTAL	13		25		30		2		13				83	

Note
For ERASMUS, Poland, Belgium, Denmark, Norway, Romania and Turkey are not listed individually because there were only 15 students in total coming from these six countries.

I was amazed to learn right from the beginning that in Germany 'the ball is in your court', so to speak. What you do or do not do at university is almost entirely up to you. This was very difficult to get used to, coming from England where my whole course was essentially pre-determined. After making my way laboriously through the well-known bureaucracy (registering with Aachen city, the university and the Mechanical Engineering Department, among others), the system then left me totally to my own devices without any personal guidance. I had to find out everything for myself, the most difficult being not only my lecture timetable, but also which exams to take, and how to revise for them. I found it very frustrating and difficult at the beginning, not having any human contact point. Now, for me this was the hardest thing to come to terms with, going from a System which cossets and supports you to a system which has no in-built support mechanisms. You and your inner resources are your only support. This confronted me with my own limitations and forced me to develop my own personal coping strategies to overcome these barriers . . . I learned to manage my own independence to a greater extent than I ever had to.

Of course, other students were able to benefit in equally valuable ways; Fig. 9.1, for example, shows an Aerospace student standing by the plane in which he obtained his pilot's licence during his year in Toulouse.

The ERASMUS programme, which began in 1987, became incorporated in the SOCRATES programme in 1995, thereby linking the three educational stages of school, university and life-long learning together. In 1997 the system of applying for specific EU contracts was replaced by 'Institutional Contracts' in which the University was required to make a composite bid for all disciplines to cover a three-year period, and in 2007 SOCRATES was itself replaced by the Life-long Learning Programme, with ERASMUS as one of its four sectorial groups.

In this developing field of European integration in higher education, what is now referred to as the 'Bologna Process' should be recorded, because of its potential effects on the Faculty's future activities and on the employability and mobility of its students. It came from a 1999 meeting in that City, of 29 Ministers of Education from Member States, who agreed to develop by 2010 a 'European Higher Education Area' for

Fig. 9.1 James Lyon, Aerospace Dept., 2002–06. An ERASMUS student in Toulouse with the machine in which he obtained his pilot's licence.

the complete range of degrees. By 2008, a new European Quality Assurance system had been set up, complete with its own register of complying universities, and it was considered likely that listing on this register would in future be a requirement for obtaining EU funding for educational programmes. The Bologna profile for Higher Education envisages an eight-year period, split into 3-2-3 years for Batchelor, Master, and Doctorate degrees. Few would disagree that eight years from school is a reasonable minimum period for a doctorate, but the detailed split appears to satisfy few of the Member State requirements for the first degree. In the UK, the Royal Society (Ref. 18) has spoken for science and engineering by emphasising that the four-year first degree should be retained for the foreseeable future as an effective response to the changing intake and changes in the subject disciplines.

9.2 Student Fees and the Funding Of Higher Education

9.2.1 Introduction

Some information was given in Chapters 3 and 4 about how higher education was funded in the early part of the twentieth century. The national government appears to have been the principal source, but with considerable contributions in Bristol from the City Council, and through the generosity of the Society of Merchant Venturers and wealthy individual sponsors. It will be remembered that the eventual creation of the University in 1909 was in some measure due to a letter to both the UCB and the Society from the Bristol City Clerk, which contained an implied promise that greater financial support would be obtained from the government if they would reconcile their differences.

In 1910, the tuition fee for the degree course was 25 guineas per annum, reduced to not less than ten guineas if the combined incomes of the parents was less than £350 p.a. Although far from trivial for those who had to pay the fee, the total amount thus obtained was only a small part of the full cost of academic salaries, provision and maintenance of buildings, libraries and other essential services, the larger part accruing from contributions by the three organisations mentioned above. From 1919, when the UGC was created, it provided the bulk of these costs. Assistance to students in the payment of fees came through an unspecified number of awards from the City of Bristol and from the counties of Gloucester, Somerset, Wiltshire and Devon, which also included a contribution towards maintenance costs, and there were seven Society of Merchant Venturers awards for fees only.

By 1938, degree course fees had increased to £42 p.a. with small additional amounts for examinations, but the available scholarships now occupied four pages in the Faculty Prospectus, and it is of current interest that the county of Gloucester were offering *loans* (my italics) for those intending to become teachers, 'the loan repayable by quarterly instalments during the four years following the University course.'

Very soon after the end of the Second World War, the University took on a national, rather than a local role, a change greatly assisted by men and women demobilised from the armed forces. Their fees and maintenance were a reward for their service, whereas other students were a charge on Local Education Authorities unless they were successful in winning competitive awards such as state scholarship or awards offered by individual universities. To be specific, a report of a working party on university awards appointed by the Minister of Education in 1948, determined that the three types of awards should be proportioned as 2/11ths each for state scholarships and university and college open awards, and 7/11ths for Local Authority awards. Ten years later the CVCP issued its thoughts to universities, which included

> The principle applied at present is that all awards are contributions towards the expense of University education, not grants to cover the whole of that expense. The balance falls to be met by the parents . . . If the state were to finance the full independence of the student the supervisory responsibility of the university would be increased to an undesirable extent.

The actual contributions were subject to a means test, and the CVCP suggested that if the level were raised, with actual amount being reduced, then universities would probably recruit more well-qualified students. Although the Robbins Report (Ref. 7) was not published until 1963, there was, in 1958, an acceptance that a major increase in the number of university students was required, and the CVCP were doing their best to stimulate action! What Robbins did do was to raise the issue of the relative decline of fee income, making the specific recommendation that 'Tuition fees should be revised so that in future they meet at least 20 per cent of current institutional expenditure.' In 1965 the CVCP again considered the fee situation, recommending to the UGC that fees should be raised from the 1969 session to a uniform £150 p.a. for all first degree subjects. In late 1966 the Department of Education and Science responded by confirming that the average course fee for home students would be £70 p.a., but £250 p.a. for those from overseas. This differentiation was strongly opposed, the CVCP making the case for a common fee of £250 for all students, a level, they said, was required if fee income was to reach the 20 per cent suggested by Robbins.

   As recorded in earlier chapters, for most of the next 20 years universities struggled against financial shortfalls of varying degrees of intensity, with no major changes to the course fee structure, until the Jarratt Report made the recommendation which is now discussed.

9.2.2 Changes in the Student Fee Structure

One of the Jarratt-inspired decisions (Ref. 14) of the Government in 1989, which might have produced parity of funding between engineering and the physical sciences, was its decision to increase the part which student tuition fee income played in its overall financial support of universities. It will be remembered that students paid no financial contribution to the fee component of their education, and that university income from public funds for this education came from two sources; the larger part directly from the UGC on the basis of student numbers and some other factors, whilst the smaller part came from the Local Education Authority (LEA), or similar body which was sponsoring the student, and in many cases also paying the maintenance award. In reality, both parts of the fee came from Government because it reimbursed the LEA for around 90 per cent of its contribution until 1990, with local rates paying the rest; from 1990 the Government reimbursed the LEA the full amount.

In 1988 the common annual fee for all university courses – from the cheapest to the most expensive – was £607, and in total this represented around 9 per cent of the total recurrent funding for higher education, which stood at £3250m. Then, as now, the funding body was not dictating to universities how the allocated amount should be used to support individual departments; as autonomous institutions they made their own decisions on this matter.

In 1989, the Government, pursuing the idea of making universities more efficient, suggested a two-stage change in the way in which universities received tuition fees. In the first stage, it proposed to increase the fee common to all courses to £1600, a number arrived at from the fact that the lowest average cost per student to public funds across all subjects – as calculated by the UGC – was £1700. By making the common fee slightly lower, the government felt that it was still leaving the UGC/UFC with some influence, even on the cheapest courses, whilst it still retained a major role in the more expensive ones. It was seen earlier that the Faculty had been willing in 1987 to take 43 'fees only' students, and since this was not an isolated occurrence nationally, the government had been allowed to take the view that by increasing the fee, it would persuade universities to take more students at less than the full cost.

It was also in 1989 that the UFC had to propose to universities a Government idea for increasing student numbers at no extra cost.

Competitive bidding for students was suggested, it being assumed that the UFC would be required to award the contract to the lowest bidder. Of the many grave dangers of such a scheme were the likelihood of it leading to some two-year degrees, coupled with a perceived lack of competence at the UGC to properly assess the merits of rival bids. Relief followed the withdrawal of this idea!

The second stage of change to the fee structure was in fee differentiation for different subjects, thereby reflecting real costs to a greater extent. Four groups – A to D – were suggested, in which annual fees were to be £1600, £2000, £2400 and £3200, respectively. Taking them in reverse order, the most expensive subjects of dentistry, medicine and veterinary science were in group D; engineering, science and the performing arts were in group C, whilst the humanities and social sciences were in group A. In group B were all other subjects. It was noted at the time that the national breakdown of total student numbers into these groups was, from A to D, 45, 20, 30 and 5 per cent, and that the transfer from the two funding councils – the UFC and the PCFC – would be around £700m if the two stages were implemented. The parentage of Jarratt in this fee differentiation plan should also be noted from its statement 'A fee structure which took account of this variation would clearly apply the desired market disciplines more widely within institutions.'

The first element of the student fee change was relatively easy to implement, and was carried out for the 1990 session, when for home and EC undergraduates it was raised to £1675, with the UFC claiming that the change was financially neutral for total public expenditure, with recurrent grant being reduced by a sum equal to the estimated fees paid by LEA. For comparison, the recommended minimum fee for an overseas student for the same year was £4770, £6360 and £11,660 in arts, sciences and clinical subjects, respectively, which more closely represented real costs.

From 1993 onwards it was becoming clear that the Government intended to require students themselves to contribute to the cost of their education, but this did not formally appear until the Government's 'Teaching and Higher Education Act' of 1998. In addition to requiring that all but the poorest students pay tuition fees, from the 1999 session maintenance grants were to be replaced by loans, which were to be repaid from the time when the graduate's income reached £15,000, and at a rate of 9 per cent of this income per year. However, knowing from his CVCP activities what was certainly in prospect, in late-1996 the Vice-Chancellor, Sir John Kingman, advised Senate that the 1998 under-graduate prospectus would include a provision to enable the University to charge a supplementary tuition fee above that normally paid by the LEA on behalf of home students. This proved to be a wise precaution because the HEFCE had been busy devising a new method of allocat-ing teaching funds, to replace the old approach which had been based on historical, rather than educational, factors. The new method accepted fee differentiation between four groups of subjects, but with some changes to those originally set out by the Government. For no rea-son which was explained, it changed the order from its earlier proposal, so that group A now contained the expensive subjects such as medicine, dentistry and veterinary science; B listed laboratory-based subjects such as science and engineering; C included subjects which involved a studio or fieldwork component, and D was everything else! In the original grouping the cost weighting between top and bottom was a factor of 2, but in the 2007 version of the HEFCE's new system, it became 4 for A, with B at 1.7, C at 1.3 and D at 1.

It is recalled that the UGC, the UFC and now the HEFCE, had always provided universities with its own calculation of the costs of tuition, but it was a total sum for all subjects, leaving individual universities to devise their own distribution system between departments. Now, the HEFCE was being explicit on how the total amount had been made up, and in reality, indicating in broad terms how universities should use it!

Taking the 2007 session as an example, the 'base-prices' – an HEFCE term – attached to groups A to D, were £15,332, £6516, £4983 and £3833, respectively. But, consistent with Government wishes, it was pol-icy to require all students to contribute financially to their education, an amount which they personally paid to their university. For 2007, this amount was set at £1225 for all subjects, and was deducted by the HEFCE from the amounts which they paid to a university for A to D as set out above.

This simple picture is complicated somewhat by a Government deci-sion in its Higher Education Act 2004 to introduce variable tuition fees from 2006 onwards up to a maximum of £3000 (plus an allowance for inflation) until at least 2010. To exercise some control over what uni-versities actually did, it established an 'Office of Fair Access' (OFFA), charged with agreeing the intentions of any university wishing to increase fees which students paid above the current level of £1225. Of course, even when added to the HEFCE 'base prices', £3000 was very far short of the annual cost of educating each student, so that in recognition of this fact, almost all universities were permitted to charge the maximum. As an example, for a new UK or EU student starting in the 2007 session, the fee in the Faculty was £3070, and for overseas students it was £13,100. For students who had entered the Faculty before October 2006, the earlier fee of £1225 was charged, with remission to zero if the student was intercalating an ERASMUS year in Europe, and to £615 if the intercalated year was in industry, or 'abroad'. It should be noted, however, that the University was able to provide means-tested bursaries, hardship funds and scholarship to support students during their period of study, and in the Faculty, many students have been able to obtain sponsorship from industry covering fees and some contribution to living expenses.

In 2008 the HEFCE began reviewing its method of allocating funds for teaching, and have already made some changes. Of some significance also, is the fact that the Government are reviewing the current £3000 (plus inflation) limit on what universities can charge each student. In a House of Lords debate in July 2008 the average annual cost of teaching an Oxford undergraduate was given as £20,000, with corresponding income at £7,500, the difference being provided by a subsidy from the research element of the HEFCE award. In the same debate, our own Vice-Chancellor, now Prof. Eric Thomas, was quoted as saying that Bristol would, as an average, need to charge an additional £19,000 p.a. for each student if teaching costs were to be fully covered. If a more realistic upper limit is to be set by the HEFCE, it could have the effect – OFFA permitting – of introducing a variability in fees between those universities which have no difficulty in attracting good students, and those that do.

9.3 The University Campaign for Resource

The previous chapters have been notable for increasing governmental control of university developments, coupled with reduced funding for an increasing number of students. Universities were exhorted to find other sources of financial support, from industry, from research councils and charities, as well as soliciting legacies and gifts from former students in the manner known to be successful in the USA. In pursuance of this need, the 1989 University Plan set out its strategic objectives, and shortly afterwards it launched a 'Development Campaign' with Dr Alisdair Lockhart appointed to be its Director, it being agreed in Senate that all future attempts by academic staff to obtain development funding should be coordinated by him. Included in the plan was a 50 per cent increase in student numbers during the next decade, to around 11,000 students, a doubling of the research activity, and a 30 per cent increase in Continuing Education. The cost of capital projects necessary to achieve this was estimated to be £100m at 1990 prices, and at the head of this list of 14 major projects was 'A New Building for Expansion of the Engineering Faculty.'

We shall see shortly that this became the new Merchant Venturers Building (MVB), although the principal contribution to it from the Campaign was in persuading several Bristol companies to rent parts of the MVB not immediately needed by the Faculty.

A powerful 'Campaign for Resource' Committee was established, with Sir Michael Angus as its chairman. During 1949–52 he had been a mathematics student in the University, and after service in the RAF progressed in industry to become Chairman of Unilever plc, and President of the CBI in 1993. The Campaign was to be project-driven, meaning that funds were to be raised for individual capital projects which would stimulate academic developments and realise the University's research potential; it was certainly not for general recurrent expenditure. This last point was important because for the session ending in July 1990, the University had found what appeared to be a deficit of £4m in its finances, to be attributed later to a lack of expertise in the Finance Office for handling the devolved budgeting process. This deficit affected the Faculty because its immediate effect was a University-wide 15 per cent cut in revenue budget and a freezing of all vacant posts. The cut, exacerbated by a £1m reduction in UGC grant, was continued into the 1991 session, together with a restriction – enforced previously during the 1980s financial crisis – that '. . . departments which are able to earn substantial income may not benefit from that income in order that the University as a whole does not go into deficit.' Fortunately the loss was quickly recovered and no lasting harm was done.

The Campaign Director's role at the practical level was to liaise with Deans of Faculties, exploring where research developments could be made, in areas where industry might prove willing to provide the financial resources, and to arrange contacts between the various partners. Three examples can be given where this process was successful in the

Faculty. In Civil Engineering the concept of 'risk' and its partner 'safety', had been a research theme initiated by Sir Alfred Pugsley (Chap. 4) and widened in scope by David Blockley, making use of new concepts in information engineering implemented on computers. The invitation to extend the boundaries of this research was accepted by four industrial partners, Nuclear Electric, Railtrack, Lloyds Register and the Civil Aviation Authority through its National Air Traffic Services. This group combined to provide the initial funding for a five-year period of support for a Director, to be seconded from Nuclear Electric, and two full-time researchers. In 1995, after early success in establishing its unique contribution to new thinking in this field, its status was formalised as 'The Safety Systems Research Centre (SSRC)'. The underlying objective of the SSRC was to address the problems highlighted by the design, assurance and operation of safe and reliable computer-based systems which are of a fundamental and philosophical nature, and with this in mind it was initially part of the Computer Science Department, but was transferred to the Civil Engineering Systems Group in 2003, by which time it was no longer supported exclusively by industry, but had obtained EPSRC funding with which to extend its research base. In 2006 its role was further strengthened by the creation of a 'Systems Performance Centre', an alliance between the University and British Energy. Its purpose is to deliver strategic research in understanding and monitoring nuclear plant and in improving operational performance and management, making use of the EngD opportunities for postgraduate education and training (Chap. 10.4.5).

A second success of the Campaign for Resource, also primarily in Civil Engineering, was in the crucially important area internationally of water resource and management. Its industrial sponsors were Bristol Water, South-West Water, Allied Domecq plc and Unilever plc who undertook a five-year commitment to

> develop and direct research in the systems approach to water management globally, promote the dissemination of information on best practice in water management and develop undergraduate and postgraduate teaching programmes in the more general area of Environmental Engineering.

The Water and Environmental Management Research Centre (WEMRC) was established in October 1997, with Prof. Ian Cluckie as its founding Director, and now has research students from many parts of the world. It is noted that of the four sponsors, the Chairman of the University Council, Moger Woolley was at that time also Chairman of Bristol Water, and Sir Michael Angus, the Campaign's Chairman, occupied the same position in Unilever. Their influence on the Faculty's behalf has been well repaid by the international status which the WEMRC has achieved.

Two other successes in the campaign, which, like the previous two, involved considerable effort by the Dean, was first the ICI Chair in Process Engineering, to be filled in 1992 by a nuclear engineer, Prof. G. L. Quarini. Apart from his research (Chap. 16) he proved to be an inspiring teacher, illustrating his lectures with physical models of his own creation. It was not surprising that in 2008 he received the University Teaching and Learning prize. Due to ICI hiving off its pharmaceutical interests to form Zeneca plc, the funding for the first five years of this appointment was provided by that firm. Earlier, in 1990, £500,000 was obtained from Toshiba Research Enterprises (Europe) Ltd. for a Chair in Communication Networks, which was filled by Prof. J. P. McGeehan from Bath University.

The Faculty also became involved in a Campaign project which stemmed from the 1993 government White Paper 'Realising Our Potential' (Ref. 17). This drew attention to the need to attract more young people into science and engineering for the benefit of the economy and quality of life in general, coupled with a better public understanding of the issues involved. In 1996 the University accepted this challenge by proposing the creation of a chair in 'The Public Understanding of Science and Technology', to be complementary to the chairs at Oxford and Imperial College having the same title, and to be funded for ten years as a memorial to a patron of the Campaign, John Collier, FRS, FREng, who had recently died. He had been Chairman of Nuclear Electric plc, based at Barnwood, near Gloucester, and had regularly invited members of the University (including the author) to working dinners at Barnwood for discussions on scientific and engineering issues with senior and influential (but non-scientific) members of the public. A half-time visiting appointment for one or two years was envisaged, based in the University's Institute for Advanced Studies, and would draw upon, and provide leadership to, existing activities in this area. For the £500,000 required, 23 sponsors had been assembled early in 1997 headed by Magnox Electric and Smith Kline Beecham, and half the required amount was immediately obtained. But by 2005 the importance of this activity became such that a full-time permanent 'Chair of Engagement in Science and Engineering' was created, which in 2006 suffered a change in name to 'The Collier Chair in Science and Society'. All through its existence, the Faculty has played a significant part in its activities, particularly in the 'National Week of Science and Technology', by exhibiting its research through displays at the Galleries Shopping Centre in Bristol.

One project in the Campaign for Resource in which the Faculty had an interest, but which did not succeed, was a suggestion in 1996 to create a university campus in Malaysia, an argument for which was that it would lead to a larger number of postgraduate Malay students wanting to carry out their research in Bristol. It was said that the Malaysian government was anxious to attract high-class UK universities to set up campuses in their country[3], particularly dealing with science, engineering and medicine, but exactly what they were prepared to offer as inducements was not clear. Initially the Deans Committee gave it qualified support, but the Faculty Dean had to explain to Senate that the great majority of his colleagues were opposed to the idea on the grounds that they were already overstretched and under-resourced. Despite this lack of enthusiasm, a small group of University officers visited Malaya, reporting back that they had identified a sponsor who would provide a site. However, wise heads prevailed, and early in 1997 the University wrote to the Ministry of Education in Malaya '. . . to confirm that it cannot proceed with its plans to establish a campus in Malaysia jointly with a Malaysian partner, but it would be willing to assist in establishing a Medical School.'

The Faculty leaders were to some extent to blame for the failure, because they had allowed the University to think that they would be willing to play a major role in this enterprise, whereas, in fact, there was little support for what would have been an onerous teaching responsibility, with uncertain financial reward.

New initiatives in the Campaign for Resource ended with the 2002 session when Dr Lockart resigned to take up a post at University College

---

[3]   In 2000, Nottingham University was the first UK university to establish a campus at Semenyih in Malaysia. By 2008 they had also established a campus at Ningbo in China, and several other UK universities had created campuses in various parts of Asia.

London. From that point forward, the University began to build up its resource-raising activities for the Centenary in 2009.

9.4 The Merchant Venturers Building

Since 1989 the University had been in discussion with the City Council about the future shape and size of the Precinct area, the result of which, in 1993, was a consolidation of property holdings in Tyndalls Park and St Michael's Hill, together with the potential for development at the lower end of Woodland Road. Here, in the triangular area formed by this road with Park Row, the University already owned the buildings at the apex, which in recent times had been used as a cinema, an aircraft factory and a garage, and was then being used by the Faculty as an overflow drawing office. The University also housed the pre-Clinical Veterinary School in a substantial area of the triangle fronting onto Park Row adjoining the Wills Memorial Building. Although there was some private property on the Woodland Road side of this triangle facing Queen's Building, it was known that the City Planning Authority would support the use of compulsory powers for the University to purchase these houses, and this was done.

Even after Geology and Mathematics had vacated Queen's Building in 1985, the continuing growth in student numbers, coupled with success in the acquisition of research funds, had caused the Faculty to utilise every possible area of space, even a few rooms behind the synagogue in Park Row! Several abortive schemes had been explored for creating substantial new space, including a tower block below Queen's Building on the site then occupied by a chemistry department car park, and to the enclosure of the Queen's Building courtyard by an atrium roof with laboratories below.

There were four principal reasons why the University chose to allocate the Woodland Road/Park Row triangle site to the Faculty. First, was the fact that the Computer Science Department had joined the Faculty in 1989. Second, following one of the recommendations of the Finniston Report (Ref. 4) that a four-year undergraduate course leading to the MEng degree should be offered to the top group of engineering students, the Faculty decided in 1992 that it must move wholly in this direction, with a consequent increase in student numbers. Third, the explosive growth of interest in electronic and communications engineering could not be contained within the confines of Queen's Building because, taking agreed expansion into account, by 1994 the Faculty would be 9200 m^2 short of the space norm calculated by the university to be 20,750 m^2 in total. The fourth reason for the University's allocation of this proposed new building to the Faculty was financial. It had probably always been true that the income/expenditure account of the Faculty had shown a positive balance in the University's accounts, and certainly from the early 1980s when it was found necessary to make such figures available, its accounts had shown large surpluses. During the 1981–4 financial crisis (Chapter 6) for example, in each session it was able to save considerably more than its required target, and following the introduction of the Resource Allocation Mechanism (RAM) it became clear that the Faculty was providing substantial support to other parts of the University (Table 8.1). In presenting the RAM, the Deputy Vice-Chancellor, Prof. Brian Pickering, explained his proposals – which the Faculty accepted – in the following terms:

> It is acknowledged that it is in the University's interest for
> some budget centres to be resourced at a higher level than
> their attributed income would suggest. Provision is made for a
> strategic redistribution of income to achieve this.

In the 1994 session, for example, this 'strategic distribution' meant that the Faculty's attributed income was greater than its attributed expenditure by more than £1m. It was this financial record, and the understandable wish that it should be further encouraged, that prompted the University to fund the new Park Row development by borrowing the £12.5m required, with the Faculty agreeing that initially one-half of the space created would be rented to commercial organisations, thereby helping to fund the amount borrowed. To allow for the forecast expansion of the Faculty, the space rented was planned to decrease over a ten-year period, so that by 2005 it would be wholly occupied by the Faculty, possibly by 'spin-off' companies which its research had created. This is now due to occur in 2009.

The question as to which parts of the Faculty were to occupy the new building raised the important issue of academic coherence. One of its principal strengths, noted by many universities and envied by most, was that all Departments occupied the same building, which made for routine daily contact by staff and students, to the benefit of both teaching and research. The obvious options for the Park Row site were to try to retain some of this coherence by splitting Departments, or, alternatively, to divide the Faculty. The second option was chosen, largely on financial grounds, with Aerospace, Civil, Mechanical and Engineering Mathematics staying in Queen's Building, and Electrical, Communications, and Computer Science occupying the new building, together with the Faculty computer teaching laboratory. A pedestrian bridge across Woodland Road, or a tunnel underneath it, was suggested as a means of continuing the easy contact between the two parts of the Faculty, but cost again ruled this out. Perhaps it was a natural consequence that growth would entail division, and this was certainly the case. One practical consequence of the split site was the necessary development of a 'student information system' on the Faculty web-site, providing information and interaction between students and academic staff

The triangular site for this new building, with large changes of level on all three sides, caused major problems for the architects. The preservation of the stone façade of the original Woodland Road/Park Row apex building was a condition imposed by the City authorities, but this had the pleasing effect of prompting the liberal use of stone facing in the new building, allowing visual continuity with surrounding buildings. One condition imposed by the University Building Committee was that it should be as energy efficient as possible, and this resulted in many novel features to be introduced in order to generate an efficient internal air-circulation system in the atrium. The distinctive upward-sweeping curved roofs were an architectural feature designed to give the effect of a 'modern tower' in keeping with the towers in neighbouring university buildings

In deciding on a name for the new building, it was recalled that the Society of Merchant Venturers had not only sponsored engineering education in Bristol since the seventeenth century, but had played a major part in creating the Faculty in 1909 (Chap. 3), and had given it very generous support until 1950. It had also recently provided the finance for 'fitting out' this new building. Accordingly, the Vice-Chancellor obtained the Society's permission to give it the name 'Merchant Venturers Building'. The 'MVB' was officially opened by HRH the Duke of Edinburgh on 19 July 1996, and followed by a Faculty party! Fig. 9.2 is a photograph of the MVB taken from the top of the Wills Memorial Building, and it can also be seen at the top-centre of Fig.10.3 (by courtesy of Google Earth).

Fig. 9.2 The new Merchant Venturers Building; a photograph taken from the tower of the Wills Memorial Building. Queen's Building library is at top left.

9.5 Undergraduate Teaching and the Royal Academy of Engineering (RAE)

From its beginning in 1660 the Royal Society of London had many members who would today be recognised as engineers, and this situation continued for more than 300 years, until in 1976 a group of its Fellows, which included four from the Faculty, met to discuss the creation of a body which would have similar objectives to the Royal Society, but would be for eminent engineers. Many members of this group were also involved with the Council of Engineering Institutions (Appendix 2), which had HRH the Duke of Edinburgh as its President. His interest in engineering, and in the contribution which it was required to make to the British economy, was reminiscent of an earlier Queen's Consort, Prince Albert (Chap. 1), and his influence was to lead in the same year (1976) to the creation of the Fellowship of Engineering. Initially, its 126 Fellows were nominees of the Professional Engineering Institutions, and they gave themselves the annual task of electing to Fellowship up to 60 of the most eminent engineers in the UK, but with an overall upper limit in due course of 1000 Fellows. At a later date this limit was eased somewhat, and the election of a small number of overseas Fellows was allowed. Its success in representing UK engineering at national and international levels, coupled with the Queen's Consort as its Senior Fellow, resulted in the Fellowship being granted a Royal Charter in 1983, and to becoming the Royal Academy of Engineering (RAE) in 1992. By this time, the RAE had become the de facto interface with Government on engineering issues and, like the Royal Society, merited substantial grant-in-aid from public funds. As its prestige, influence and income quickly grew, in 2007 it was able to move its headquarters so as to be adjacent to the Royal Society in Carlton House Terrace.

A theme running through all the RAE's activities involving universities is that of providing links between them and industry. At one level, in 2008 there were three schemes for providing Visiting Professors (VP), and at another there were awards for the most promising engineering undergraduates. It would not be universally accepted, however, that the RAE's involvement with university education exhibited a complete understanding on their part of the different roles which universities on the one hand, and industry on the other, should be called upon to play in the 'formation' – an Engineering Council term – of the professional engineer; and whereas there is no doubt as to its good intentions, its effects have sometimes been to swing the delicate balance between teaching and research in universities in favour of the former activity, and even to attempt to introduce material which is closer to training than to education.

Visiting Professors in the Principles of Engineering Design
In 1991 the Faculty was successful in being able to appoint four VPs in this scheme, which was supported financially, through the RAE, by Charter Consolidated plc. The four original appointments were Roland Bertudo, Director of Strategic Planning for the Rover Group, Chris Elliott, Director of Smith Systems Engineering, Michael Shears, Director of Ove Arup, and Ted Talbot, Chief Engineer, Airbus Support

at BAe Airbus. The first and last of this group retired in 1994 to be replaced by Jeremy Davies of the Computer Engineering Group at Rover, and Horst Peter, Managing Director of ESAB Cutting Systems at Farben in Germany.

The early contribution of these VPs to the Faculty was through undergraduate lectures and workshops, presenting their different interpretations of the design problem, and in this their success was such that the Faculty began, with the help of industry contacts, to develop a five-year MEng course in Engineering Design, which involved one year with one of the 12 major companies which had agreed to support the course, as well as placements with these firms during the summer vacations. A novel, and apparently very attractive feature of this course, is the opportunity to select 'open units' (Chap. 8.11), equivalent to one-third of a year's study, from anywhere in the University, subject, of course, to them being consonant with the educational development of the student – as determined by the University and industrial tutors. In the Faculty the principal sponsoring Department for this MEng was Engineering Mathematics, which offered it for the first time in 2001, having obtained professional accreditation for chartered status with all the major Engineering Institutions. Locating it in this Department ensured that there would be no bias towards any particular branch of engineering; indeed, it may be considered as a reversion in some measure to the much-respected engineering–science type of course originally offered by the Faculty during its early years, but at the higher MEng level.

This VP scheme in the Principles of Engineering Design was taken up in due course by 44 UK universities, resulting in 140 VPs being appointed by 1997, after which new RAE funding was directed towards the development of the following two schemes.

Visiting Professors in Engineering Design for Sustainable Development
During 1998–2004 the RAE sponsored 26 universities to develop teaching materials, based on case studies, with which to enhance the understanding of sustainable development, embedding them in a teaching and learning culture. This activity is based on industrial practitioners working as Visiting Professors for about one day per week alongside university teachers experienced in the art of delivery to students. In 2007 the Faculty was successful in obtaining a £60,000 award in this scheme, and to appoint as VP for five years, David Welch, the director at Rolls Royce responsible for Health, Safety and the Environment. His

remit is to develop teaching material on the theme 'Power Generation in Developed and Developing Countries'.

Visiting Professors in Integrated System Design
The third, and most recent, of the RAE's initiatives for linking industry and universities at the MEng undergraduate level is based upon a belief that the syllabus will be enhanced, without compromising intellectual rigour, if students can be given an 'understanding of the importance of systems thinking and a whole product holistic appreciation, before being limited by the confines of a narrow field of study.' The last phrase here is unfortunate; it would be argued by the majority of university teachers, including those in engineering, that the essential purpose of university study is to give the student the opportunity, and time, to master difficult concepts through study in depth. Breadth comes later, and is the responsibility of the training programmes of the Engineering Institutions, possibly in conjunction with the postgraduate EngD degree promoted by the EPSRC, which is described in Chap. 10.4.

In 2007 the Faculty obtained RAE funding for three VPs in this programme, with the task of providing such material on 'systems engineering' as may profitably be fitted into the Faculty's undergraduate programme.

Engineering Leadership Advanced Award Scheme for Undergraduates
Known originally in 1995 as the 'Engineering Leadership Awards', this scheme provides support and motivation for the very best engineering undergraduates in UK universities. In the early years, 25 such awards – each up to £7500 in value – were awarded every session to provide MEng undergraduates with business training and experience during the vacations, or immediately after their university course. Later on, the number was increased to 30 each year and the value reduced to £5000. The competition for these awards is extremely intense, involving psychometric testing, essay writing and interviews during a weekend residential assessment. But Faculty students have done very well; during the thirteen sessions 1996–2008, a total of 29 students have been successful, which is a remarkable achievement bearing in mind the number of universities competing. Of this number, there were seven each from Electrical and Civil Engineering, and five each from the other three departments, Computer Science not being involved.

9.6 Teaching Quality Assessment (TQA)

Since the middle of the nineteenth century, the quality of the under-graduate teaching in UK universities has been assessed by the Professional Engineering Institutions (PEI) through their own examining boards which visit universities on a five-year cycle. In addition, each Department has had its own external examiners for validating the quality and probity of its teaching and examinations in all subjects. On only one occasion – to the author's knowledge – has any Department of the Faculty been even mildly criticised by the PEI, and that was in 1986 when the IMechE considered that the MEng/BEng borderline criteria were not sufficiently clear!

But in 1993 the HEFCE – responding to government pressure – decided that it would itself demonstrate the quality of university teaching by an external audit which imposed a major additional administrative load on all categories of staff. The two-day visits to each department by assessors who had offered themselves for this task, and who were rewarded financially for doing so, required the production of lecture and laboratory notes, examination papers and marked scripts, and other material of such volume that it would be surprising if it was thoroughly read and assessed. The six areas to be considered in this assessment of teaching quality were:

> Curriculum Design, Content and Organisation
> Teaching, Learning and Assessment
>
> Student Progress and Achievement
> Student Support and Guidance
>
> Learning Resources
> Quality Assurance and Enhancement.

What generally concerned academic staff was the attendance of the assessors at lectures and laboratory classes. Too often they proved to have stereotype ideas as to the manner of delivery, criticising any personal approaches; only seldom did they appreciate that fine teaching is an art, in which the idiosyncrasies of the teacher plays an important part. Despite the lack of enthusiasm for the activity, the Faculty came through these TQAs very well. Until 1995 the assessment results were recorded in three categories – excellent, satisfactory and unsatisfactory – and in 1993 Mechanical Engineering, the only department to be assessed under this system, was considered to be 'excellent'. From 1995 the actual scores were given for each of the six areas on a scale of 1 to 4, and for four departments, Civil and Aerospace received 22 points, Engineering Mathematics 23, and Electrical 24. For the first two Departments the apparent deficiency was in each of the last two areas above, and in Engineering Mathematics it was in the fifth. No doubt Faculty staff had emphasised to the assessors the information with which to make this criticism of learning resources, because they were well aware that much of their laboratory equipment was old and inferior to that which the assessors would have found in the newer universities. That this situation was in the process of being remedied by the HEFCE itself in its 1997 Joint Research Equipment Initiative (JREI), would seem to indicate that by making their criticisms, the TQA assessors were actually trying to help the Faculty! This did prove to be so, because from the HEFCE initiative the Faculty received £918,400, made up of two awards; one to Prof. McGeehan (Electrical) for an 'Integrated Communications and Computing Laboratory', and one to Dr Taylor (Civil) for equipment for the earthquake engineering laboratory. When the JREI was repeated in 1998, Prof. David May was awarded £362,000 for a 'Bristol Image and Video Archive'. The Research Councils also held a competition for the provision of research equipment, and in this Dr Colin Campbell (Eng.Maths.) succeeded in obtaining £82,000 for the purchase of high-performance workstations for computationally intensive research projects in Aerospace and Mechanical Engineering.

Unlike the Research Assessment Exercises discussed in Chapter 8.9, the gradings obtained in the TQA did not directly affect the HEFCE allocation of funds; but because they became public knowledge, they were thought to have had an indirect consequence on the choice of university made by schoolchildren, and therefore on university income derived from tuition fees.

TQA was clearly considered by Government and the HEFCE to be a necessary activity, because in 1997 the Quality Assurance Agency (QAA) was established, funded by the latter body, together with subscriptions from the universities and colleges of higher education. Its function was to QA each institution as a whole, assessing whether it had systems in place for it to QA its own departments. The University was successfully subjected to this process in 2004, and then began its own system of internal departmental reviews, with a panel made up of a Pro-Vice-Chancellor, the Faculty Dean, two external assessors, and a number of internal members including the Head of Department. The University body to which these panels report is the Planning and Resources Committee (UPARC), and in 2004 and 2005 Civil and Aerospace were assessed, both with satisfactory outcomes; Mechanical followed with similar success in 2008.

The important report by Sir Ron Dearing is discussed in the next chapter, but we note here that one (No. 25) of its many recommendations – which was not implemented – was that, as a means of attempting to ensure teaching quality, universities should lose the opportunity to choose their own external examiners; they would instead be selected from a pool which the QAA would appoint.

The disquiet over the TQA and QAA was not confined to the Faculty. It was considered by many that a Vice-Chancellor who is properly supported by a group of senior colleagues would know which were the weak/strong departments, and would know how to deal with them in the best interests of the university. Even so, it was accepted – with regret – that the issue was not an internal matter, but a question of satisfying the paymasters that their money was being properly used. However, the concern over academic freedom was very real, as is indicated by the following written question for the Vice-Chancellor at a Senate meeting:

> Does the Vice-Chancellor believe that the University is
> satisfactorily discharging its obligation to its staff to ensure
> their academic freedom in the face of threats from
> Government and its agencies and the audit culture in general
> both without and within (academic freedom entailing the
> right of every scholar to persue and promulgate her/his view
> of the truth in what s/he believes is the most effective way, and
> not to have to bow to prevailing orthodoxies and ideologies,
> however fashionable or powerful)?

Whilst expressing a measured degree of sympathy with the questioner's concern, the Vice-Chancellor replied that secession from the QAA would be illegal and that it was important to make every effort to influence the debate on quality assurance issues in a constructive way.

9.7 The University and Faculty Computing Facilities

9.7.1 University Facilities

The first expressions of interest at University level in digital computing appeared in 1959, when Senate formed a Computer Committee consisting of Professors from the Science and Engineering Faculties, which reported that it 'unanimously recommended the provision of an electronic computer within the University'.

It had been prompted to do so by younger colleagues in these two Faculties who had been using machines such as Pegasus at Southanpton and Deuce at BAC Filton. The machine proposed by the committee was the Ferranti Orion, to be purchased for £90,000 and located in the new Chemistry building. To manage and operate the machine, the committee proposed the establishment of a computer unit as a sub-department of Mathematics, with Dr M. H. Rogers as its leader. Looking ahead, it was not until 1971, that it was designated as a Computer Centre, with Prof. Rogers, as its Director and also as Professor of Computer Science in the Department of Mathematics.

In refusing the funding for the Orion, the UGC offered a second-hand Ferranti Mercury machine from the University of London, which was about to be replaced by the more powerful Atlas computer. But London decided that it wanted to retain the Mercury whilst undertaking the not insignificant task – 50 years ago – of getting the new Atlas machine to work satisfactorily. In 1963 the University used the £12,000, which it had set aside to maintain the Mercury, in buying a small IBM 1620 machine, supplemented the following year by a larger one, an Elliott 503 and both machines were soon running at full capacity.

By 1967 Government-funded Computer Board had been given the task of providing universities and Research Councils with computers, and, having received individual requests from the universities in the South-West, it suggested that by networking they would each have access to greater computer power. The working party which took this suggestion forward consisted of representatives from the universities of Bath, Bristol and Exeter, and the Welsh Colleges. After assessing computers then available from British manufacturers – a condition imposed by the Board – they decided on English Electric System 4 machines, and it was agreed that System 4-50s would be installed at Cardiff and Exeter immediately, that Bath would receive the same machine in 1969, and Bristol the more powerful 4-75, to be installed in the new Mathematics building. The growing importance of computing was signalled in 1968 by the appointment of the Pro-Vice-Chancellor and Head of the Aeronautical Department, Prof. Collar, as Chairman of the Computer Committee.

Although the idea of networking computers was not new, there were many practical problems to be overcome, so that in 1970, ICL and the five institutions in the proposed SW Network agreed to work together for three years to achieve software for this purpose. In 1971 the Computer Board visited the University, giving it the opportunity to explain the difficulties it had in using the 4-75, and to comment that universities should not be expected to develop a computer system simultaneously with providing a service for teaching and research. As a result of these difficulties there was a growing unsatisfied demand in the University for computing facilities, which was only marginally relieved through a link to the London CDC 7600.

As the Computer Centre grew, so did the general dissatisfaction with the regular failure of the 4-75 machine, so much so that in May 1972 a Computer Services Review Committee was created in the University, made up of the Vice-Chancellor – now the physicist Alec Merrison – the Chairman of the Computer Committee, which from 1970 had been Dr B. M. Bird, Reader in Electrical Engineering, the Professor of Computer Science, and a representative of the users. It took as its task

> To enquire into both the present provision of computer
> services and the University's future requirements and in its
> review to include the internal management structure of the
> Computer centre, its role in relation to the Computer
> Committee and its associated bodies and to other universities
> in the South-West network.

It invited the Directors of the Computer Centres at Cardiff and Southampton to assist it in its task, resulting in the decision that the Computer Centre should be separated from the computer science section of the Mathematics department, and should have a full-time director. This was the beginning of the realisation that the Computer Centre of the future would provide a service to the whole University, and that Computer Science was not exclusively a sub-section of Mathematics. Indeed, in 1989 it found its Faculty home in Engineering and its national home – through the British Computer Society – in the Engineering Council.

Returning to the SW Computer Network, its first software was released in October 1972 as a result of the agreement with ICL, and it was estimated that user trials would take as long as a year due to the need for each of the four machines to be introduced separately. The simple language BASIC was to be available on all machines, but more complex languages were allocated to individual machines, ALGOLW being Bristol's responsibility.

In 1973 Dr Bird became Chairman of the SW Network, and in 1979 Bristol was designated as the location for a new large computer serving this network. But since it was still required to be a British-made machine, the University expressed dismay at again being required to rid the machine of its design faults, whilst providing a service to users. The only consequence of this disquiet was the Computer Board's decision that a SW Universities Computer Centre (SWUCC) should be set up on a site in Bath under its own Director, with all costs of the new machine – an ICL 2980 – being paid by the Board until the end of 1977, after which the regional institutions would be required to cover the costs.

In Bristol the demands on the Computer Centre were growing at an increasing rate, the only solution for which was to ask the Computer Board to supply a range of machines costing a total of £1.78m over a five-year period, starting in 1975. In addition, seven new operatives would be needed, as would a new building of twice the current accommodation. Responding to this need, the Board, through the SW network, allocated £700,000 to Bristol in its five-year plan for the replacement of the 4-75.

The rapid growth of computing in 1975 pointed to the need to review its organisation within the University; it was no longer a half-time task for the Professor of Computer Science, but needed a full-time Director and this role was given to Prof. Rogers. When, in 1984, Computer Science separated from Mathematics to become a university department in its own right, Rogers became its head, and Alan Grant took on the role as Director of the Computer Centre. Also as a consequence of rapid growth, it was found necessary to review the procedure in the University for buying computer accessories; henceforth, in the

interests of standardisation and conformability, requests from departments for computing peripherals needed the approval of the Computer
Committee, and it was decreed that 'Computer facilities within the
university should continue to be centrally controlled, and that geographical location of equipment in a particular department should not
imply that use of it was restricted to that department.'

There were three main computing needs proposed for the 1977–82
quinquennium; a machine exclusively for teaching, data capture and
processing equipment for laboratories, and a hardware support group
to develop the use of microprocessors. All these depended on University
funding, and all received support; in particular, a Prime 300 machine
and four graphics terminals were obtained for teaching. By 1977 the
Computer Board had increased its allocation to the University for the
replacement of the 4-75 from £700,000 to £1m, and additionally to Bath
University £300,000, so that the two universities could form an Avon
Universities Computer Centre (AUCC) with a Honeywell machine having the 'Multics' operating system. To this machine, a number of terminals were to be attached, using a University funding of £127,000,
initiating a serious debate about whether this should be spent on
'proven', but un-intelligent terminals (i.e. VDUs), or the 'intelligent'
type which could be foreseen, but not yet available. Although the first
option was taken, modest experiments were accepted, in linking microprocessors to Multics, and the possibility of word processing. For the
first of these, the University Computer Centre made such excellent
progress that it was given a special grant by the Computer Board to
act as a distributor to other universities of a microprocessor software
development package having the acronym MICROSIM.

The introduction of the Honeywell machine into what had previously been a group of sites operating System 4 machines, was the beginning of the end of the SW Network. With effect from September 1981,
following an abrupt decline in use of the SWUCC facilities, the
University decided to withdraw from its resource-sharing arrangements. The future of the SWUCC was now in doubt, with discussions
taking place about the use of its staff and facilities if the ageing ICL 2980
was not to be replaced in 1986. In fact, the Board decided in 1983 to
provide the SWUCC with yet another ICL machine – the pre-produc-
tion 'Estriel' – to run in parallel with the 2980 during the period 1986–8.

But all was not well with the Bristol/Bath Honeywell machine either!
In 1980 its poor performance caused an exchange of letters between the

Vice-Chancellor and the Honeywell Director, resulting in a free provision of additional memory.

Although research continued to be the main driver of computing needs, the value of computers in the Library, Finance Office and other administrative sectors caused the University in 1984 to consider its future needs in such activities following the introduction into the Registrar's office of an interactive system based on a Hewlett-Packard 3000/40 machine, which had terminals in each Faculty office and in each of the examinations, accommodation and enquiry offices. It was also considered necessary at this time to revise the structure of the Computer Committee, splitting its responsibilities amongst three sub-committees dealing with Academic Computing, Administrative Computing and Technical Advisory matters. In this structure, the Director of the Computer Centre was given the new title of 'Director of Computing', the significance being that he was now in overall charge of computing within the University, as distinct from running the Computer Centre. His first task was to lead the University through new legal requirements in the use of computer facilities, as required by 'The Copyright (Computer Systems) Amendment Bill' which required him to cover possible problems introduced by at least 400 microsystem users. It is unlikely that the Faculty was of much concern to the Director in this legal problem, because its new microelectronics laboratory was provided with a VAX 11/750 for teaching and research, replacing a PDP 11/44 which was moved to Geography.

In the Centre itself, the Multics system, having received a one-third increase in power, was now working well, and in addition to having a number of new software packages for computer-aided design and algebraic manipulation, had also been used in the AUCC to develop high-level network software which was being used in five other sites under license agreements with Honeywell.

What resources the University could make available were used to buy small machines such as the VAX 11/750 for teaching purposes, two of which the Faculty obtained in 1986, so that when the Computer Board offered £1.33m for the replacement of Multics in 1988, the decision had effectively already been taken that it should be used primarily for research needs. By this time it was not feasible for the Board to require purchase from a UK computer company, thus allowing the University to decide that an IBM 3090-150E was its choice, but the fact that it was not the lowest tender meant that approval from both the Board and the

University Finance Committee was required. That these approvals were obtained was no doubt in some measure due to IBM being willing to upgrade its specification to a 3090-150S, a machine which had used a newly developed series of processors in producing a 15 per cent increase in speed over the original specification.

In addition to its promised new mainframe computer, in 1987 the Computer Centre became the 'University of Bristol Computing Service' – and at the same time it received planning approval for a new home, to be built alongside the University Library in Tyndall Avenue.

The growth in computing needs in UK universities had its effect on organisational structures. It will be recalled that at the beginning of this period the Computer Board had required five institutions in the south-west to collaborate in shared ICL machines, but in 1979 the Honeywell machine was shared by Bristol and Bath only. By 1987 the computing needs of each university warranted separate, and much larger, machines which meant that a joint management structure was not appropriate; the AUCC was therefore disbanded in 1989, with its staff being merged into the Bristol Centre. In contrast, however, there was a need for all UK universities to act collectively through the Inter-University Committee on Computing (IUCC), on actions required to implement the specific measures required by the 'Computer Misuse Act' which had become law in 1990.

In the University, financial resources could not keep pace with demand, which was felt particularly in computers for administrative purposes. In research, the shortfall was being covered by purchase from grants, and in teaching, staff and students were able to take advantage of commercial discounts in buying personal computers.

By 1992 the unit cost of processing power had been reduced to such a level that the University could see the end of the large mainframe machine, but it was not until 1993 that it made the commitment to close down the IBM service. It also decided that it would no longer purchase computers from central funds, except for 'strategic' items needed by the Computer Service for pump-priming new initiatives and maintaining the communications network; departments were free to purchase their own machines, subject to the agreement of the Technical Advisory Sub-Committee. It will be recalled that the University was now operating its Resource Allocation Mechanism (RAM), so that Faculties and Departments had greater financial freedom.

9.7.2. Faculty Computing

Computation has always been an integral part of engineering activities, with slide-rules and other devices enabling simple calculations to be made until the mechanical calculating machines were invented. Such machines allowed the engineer to hand over this work to an assistant – as was recorded in Chapter 4 when Prof. Pippard was provided with a lady computor to help with his research. By the mid-1950s the Faculty employed three ladies[4] as computing assistants, equipped initially with Brunviga machines which worked through a system of hand-operated levers, to be replaced later by the greatly superior Facit machine which had a typewriter style keyboard with hand-cranking – forward to add, and backwards to subtract! In this period many types of engineering calculation involved finite-difference methods, in which the differential equations governing the behaviour of systems were replaced by numerical approximations leading to large arrays of simultaneous algebraic equations. It was these that the ladies were employed to solve, either by an approach which involved making successive adjustments to an educated guess at the correct solution (made by the researcher) until necessary conditions were satisfied, or by direct inversion of a matrix defining a small number of variables. Although Roderick Collar had led the way towards the matrix formulation of engineering problems, it was not until the digital computer provided a means of inverting the large-order matrices, that his methods could be applied to real engineering problems.

Aside from the Faculty's major participation in the University's development of computing, it had its own activities, particularly in its teaching and administrative functions managed by the Faculty Computer Committee (FCC). In 1982 a designated computer room in Queen's Building housed a VAX 11/750 machine using the Unix operating system. Over the next decade other computers were added, most of which came from research contracts, and three terminal rooms were set up for users to access these resources. At about the same time the BBC Micro became available, which with an additional piece of software allowed it to emulate a computer terminal. Despite their cost, many Faculty members bought their own machines and used them interchangeably as terminals and as stand-alone microcomputers.

---

[4] Eileen Watkins, Christine Faithful and Mrs Tilly.

The possibilities offered by the IBM personal computer led the Faculty to a collaborative project in 1988 with IBM, in which 27 of their PS/2 machines were managed in Queen's Building by a Computer Teaching Officer, whose task was to network these machines for use in the running of an introductory computer course for all engineering students. As a result of this experience, a year later the Faculty accepted an invitation from the Computer Centre to take part in a new networking technology described as the 'Ethernet'.

As previously noted, in 1992 the University stopped 'top slicing' its income for the support of computer facilities, but the Faculty – as a University cost centre – continued to do so for teaching and general infrastructure, most notably the Ethernet which soon became its most essential component. The FCC took on the role of the overseer of the Faculty's computer infrastructure, encouraging not only the early introduction of appropriate technology, but also common standards. These became more important as each department, growing in size, appointed a computer officer to take control of its own resources.

In 1993, the Faculty and its departments were able to utilise a World Wide Web server – a piece of software which soon made changes to teaching and research (and to everyday life) which are difficult to exaggerate.

Another development at this time which affected teaching – though to a much smaller extent – was the programmable hand-held calculator, some of which had graphical displays. These became so powerful that new examination regulations were required to control their use, making a specific list of calculators which were allowed.

As computing for administrative purposes developed in the University, so it was in the Faculty, when in 1997 it appointed a Faculty Computer Officer with special responsibility for Administrative Data Support. This has been mentioned in Chap. 8.11 when discussing 'Modularisation' – a development which could not have been successfully accomplished without it – and this was also true for keeping examination and other data relating to each individual in an increasing number of students.

By the mid-1990s the Faculty had two computer teaching laboratories with 50 PCs between them, and a Unix workstation laboratory for 20 students. An expansion of this provision was made possible by the new space in the Merchant Venturers Building (MVB), which had two 50-seat computer teaching laboratories, one allocated to PCs, and the other to the Unix workstation. The MVB was wired with new Ethernet technology, and its machine room equipped with a new server which gave students their own work-space, instead of having to rely on floppy disks. As the demand for computer teaching facilities continued to grow, the old network in Queen's Building was modernised - allowing the Unix workstation to be re-located there, with a consequent expansion of PC provision in the MVB. By 2007 the Faculty had more than 200 PCs available to students, divided equally between the two buildings.

Two further contributions to computer teaching were made during 1995–8 through the development by Jon Sims-Williams and Mike Barry in Engineering Mathematics, of a 'Test and Learn (TAL)' system of computerised assessment, made necessary by the increasing number of students being taught by the same number of staff. An initial grant of £50,000 from the HEFCE/JISC, held jointly with the School of Chemistry, enabled a database of multiple-choice questions in first-year mathematics to be written. Several hundred questions were presented and classified within a menu, allowing a course tutor to set a test according to a specified prescription. Most of the questions had a correct answer – but with three well-chosen 'distractors'! The experience of student performance in repeated testing enabled difficulty factors to be attached to the questions. The TAL project received further support from a University fund which had been set aside 'to support teaching and learning excellence', and also €15,700 from the EU Leonardo da Vinci programme to develop this work with other European universities in Italy, France, Spain and Hungary.

The second Faculty contribution in the use of computers for teaching purposes was that by Dr John Davis in providing CALWARE – Computer Assisted Learning in Water Resources, developed over a four-year period under an EC-COMETT award. This software has a modular structure including animated diagrams described in audio along with digitised video. Its interactive nature forces the user to make decisions which help retention and understanding, and the interactive example sheets force the user to work through problems before being allowed access to the solutions.

Whereas TAL was a support venture for lectures and tutorial in mathematics, and was an undoubted success, CALWARE was intended as a large-scale replacement strategy for lectures and conventional tutorials. In 1994, eighty Civil Engineering and Engineering Mathematics students had their lectures replaced by computer laboratory sessions using CALWARE, followed by group discussion. This experiment appeared to be successful, in that the average examination result improved by 3 per cent over the previous year, but this was misleading because the students made known their difficulty in adjusting to the concept of learning from the computer, instead of the face-to-face experience of listening to a lecturer or tutor. Clearly, insufficient attention had been paid to the psychology of 'e-learning', requiring more investigation in such aspects before the experiment could be repeated. It is of interest to note that although CALWARE was not part of it, the UK Teaching and Learning Technology Programme (TLIP) – which had 76 projects funded by the HEFCE at a total cost of £33m – came to the same conclusion, and only a small number of its projects were successful.

Another software development initiated by the Faculty came from a £20,000 award by the EPSRC to Jon Sims-Williams, to work with Cambridge University on risks which engineering companies face during the process of tendering for contracts. Unfortunately, after a promising start, the research assistant resigned, and the work was not completed, although several of the participating companies benefited from the experience.

Like so many recent initiatives in Higher Education, the HEFCE activity in 'e-learning' originated in 'The Dearing Report' which is discussed in the next chapter. Its Recommendation 15 reads

> . . . facilitate discussion between all relevant interest groups on promoting the development of computer-based materials to provide common units or modules particularly for the early undergraduate years

and since its publication in 1997 a large amount of software has been produced with the aim of enhancing student performance, but it is only recently that universities have begun to show gains in efficiency, particularly in on-line assessment. Because of his original work in this field, in 2007 John Davis was appointed to be the University's 'Academic Director of e-learning' within the TLIP programme from which the University had received £443,000, with responsibility for coordinating all relevant activities, and to represent it regionally, nationally and internationally on matters concerning 'e-learning'.

9.8 The Faculty Library

The library of the Faculty came into being through the transfer of books and other material from UCB to the MVTC in April 1910, and, as we have seen (Chap. 4), the haphazard way in which the transfer was effected was a source of friction between Prof. Wertheimer and the UCB Registrar.

In the following year, amalgamation of library stock took place in a newly furnished library of the Unity Street building, where it was to remain until 1955. The larger share of this combined stock, of about 2000 volumes came from the MVTC, but consisted basically of works in science, craft subjects and topics in the curriculum of its 16–18 year-old classes. It was over-burdened by historical source material, and contemporary standard editions of the works of some English poets, presented to the MVTC, as made clear from bookplate evidence, by the universities of Oxford and Cambridge. Yearly additions, detailed in the Calendars of the time, were for similar instructional purpose and, if the same criteria for purchase operated then, as it did in 1949[5], they were only to be made – virtually volume by volume – as authorised by the College Registrar on financial grounds, no matter what staff recommendations supported them. Nevertheless, some journal subscriptions were already in force. The 'Minutes of Proceedings of the Institution of Civil Engineers' from their beginning in 1837 occupied five shelves, which also housed the 'Proceedings of the Institution of Mechanical Engineers' from 1847. There was a run of 'The Electrician', begun in 1890 and of 'Engineering' and 'Nature' from the same date. At least the presence of these last three titles suggest an initiative from Wertheimer and his colleagues. The only works of research importance which the MVTC contributed to the new Faculty library were the collected papers of Lord Kelvin and the electrical researches of Nikola Tesla.

The smaller, though still significant, collection of books received from the UCB for the new Faculty in 1910, was 390 volumes which included textbooks in English, as well as sixty volumes in French, and fifty in German, marked on a list prepared at the time as belonging to the 'Exley Collection': thus identifying their former owner as John Thompson Exley (1815–99), who for a brief period in 1840 to 1841 was

---

[5] Jeffrey Spittal, Faculty Librarian 1949-84, provided much of the material for this section of the Faculty's history, with additions by the current Library staff.

Vice-Principal of the short-lived Bristol College, and later, as his father Thomas Exley (1774–1855) had been before him, a Cotham school-master. The purchases of both father and son came to the University upon the death of the latter, and much of Thomas Exley's part went to the mathematics library. However, the acquisition by the Faculty of such works as Peter Barlow on the strength of materials, John Bourne on the steam engine and Luigi Cremona and others on graphical statics, all reflected close awareness of significant nineteenth century engineering developments. Some later additions, included F. W. Lanchester's 'Aerial Flight' in two volumes (1907–08) and a work on the possible construction of a Channel Tunnel (1907).

However, none of the material in the new library could be readily borrowed for, as was virtually uniform practice in all non-public libraries at that time, it was not so much their exploitation for teaching and research that mattered, as of safeguarding them against loss. This museum-orientated outlook is well reflected in the restrictive library regulations in force in Unity Street at that time. Any student wishing to consult a book had to give twenty-four hours' notice in writing of his need, and then use of the book was allowed only during library open-ing hours. Only the Governors and the Principal could actually borrow from the library for a longer uninterrupted period, and research stu-dents could only borrow material which the Principal had determined to be 'bearing on the subject under investigation'. The Librarian was to be 'appointed from time to time by the Governors', an appointment procedure which seems to indicate that the post was occupied on a short-term basis by a staff member not fully employed on their princi-pal duties. Up to 1911, for example, library supervision fell, in turn, to two lecturers in shorthand, a man who was both Registrar *and* Librarian and Anna Maria Westphal, Assistant Lecturer in German. The longest serving Librarian after the First World War was Muriel Goodall, appointed as Assistant Librarian in 1919, then Librarian in 1929, before retiring in 1944. However, even this indifferent pre-war succession in Unity Street was better than the conditions in the University at the top of Park Street, where there was neither a full-time Librarian nor any separate library building until 1923 and, then as now, no central science library. This lack of a Librarian meant that engineering books sent to Unity Street in 1910 would have been from unsupervised depart-mental locations, and would help to explain and excuse Wertheimer's testiness in dealing with the University Registrar in this matter.

A time-consuming activity which library staff contributed to, and which continued right up to 1955, was MVTC clerical work, located in the library and using some of its very limited space. Inevitably, this took an increasing amount of available resources, reducing the effort which could be spent on the needs of library users, and on the maintenance and enhancement of the collection. Shortly before the administrative separation of the Faculty from the Society's control took effect, the first Engineering Faculty Librarian, John Hoskin Lamble was appointed in 1948, but he left within a few months, and Jeffrey Spittal took up the post on 1 April 1949.

The future development of the Faculty collections and their management was Spittal's immediate concern, requiring him to make excursions to libraries in Cambridge, Manchester, Birmingham and London, as well as to the new Military College of Science at Shrivenham; in fact, searching anywhere where examples of 'best practice' might be found. There was no shortage of money, either in hand or in prospect – 'anything reasonable but no gold-plated door handles' – summarised the Dean's instructions; neither was there any lack of enthusiastic support from the newly appointed Professors, who had known large and efficient libraries in academic institutions or in major government research establishments.

At Unity Street, since there was no separate Librarian's office, little was possible beyond the re-writing of catalogue cards. But this actually allowed for the sweeping away of such reminiscent annotations as 'lost in the attic' or 'lost in the fire 1906' and the almost ubiquitous 'say five shillings', indicating that valuation had been considered as necessary a part of stocktaking as the census of books in stock. All these comments users of the catalogue had long found irritatingly irrelevant to their needs. What was more necessary was to consider, at least in outline, the construction of a subject-index to the catalogue, and that was begun, along with an onslaught on the binding of periodicals, a duty which had been neglected for many years.

In 1955 some 8000 volumes were moved into temporary accommodation on the top floor of what was to become Queen's Building, and in September of that year Janet Webber, from the Loughborough School of Librarianship, was appointed as Library Assistant. An invitation to design an issue counter for the future library was accepted and she entered enthusiastically into much of the day-to-day work of the library. Jeffrey Spittal, now finally provided with an office of his own, worked on the development of a scheme of classification, grouping books on the shelves according to their subject content, so demonstrating as obsolete the use of fixed shelf marks to identify their position. This was the first such scheme to be put into use in any part of the University system, and remained until replaced by the Library of Congress scheme in 1987.

Inter-library loans could now be dealt with 'on site'. The range of reference works and relevant bibliographies was steadily enlarged and some new periodical subscriptions were taken. New academic staff had their own requirements and were able to make clear what gaps there were in the stock. Prof. Pugsley's research activities in surface ships and submarines justified the purchase of a run of the 'Transactions of the Royal Institution of Naval Architects' and library support for aeronautical engineering was fortified by the acquisition of a complete set of reports of the National Advisory Committee for Aeronautics, transferred from the flying-boat research establishment at Felixstowe when it was closed down. Exchange schemes for similar material were also begun with national research centres in other countries. This was a time when a scheme of offering books and journals discarded from existing libraries was in operation. Lists drawn up by the British National Book Centre in London were scrutinised, and some items of historical importance and commercial value were acquired free of cost, to be added to a small existing 'special collection'.

The Faculty Library (Fig. 5.4) was opened by Queen Elizabeth II on 5 December 1958, although put into operational use a year before. It provided space for 15,000 volumes at ground floor level and seats for eighty readers. A surrounding gallery held a similar number of books in its reading bays, with space for 2300 more on shelves around the walls, and the stack-room beneath the hydraulics research laboratory could take another 45,000. The capacity of the library was expected to be reached in 30 years, but two developments shortened this perspective; more journals were taken, and space had to be allocated for the libraries of the Departments of Geology and Mathematics.

To provide for greater interaction between library staff and users, a Library Liaison Committee was formed in 1972, and its first two chairmen were Professors in the Faculty, with Jeffrey Spittal as its secretary. It worked well, and benefited from increased liaison with the Faculty when it was chaired, from 1998 onwards, by the Assistant Dean (Information Services).

Upon the appointment of Jennifer Scherr in 1971, from Twickenham College of Technology, as Spittal's deputy, accelerated progress in the development of library services could be made. In particular, an 'Enquiry Desk' was put into full commission on the ground floor, and surrounding it were the requisite volumes of abstracts, bibliographies and reference works, so removing the responsibility for reference enquiries from the counter staff. This activity had hindered the work of handling issues – at this time being manually controlled – and used three-part handwritten paper slips, which had to be 'edited' before being filed by anyone on evening duty. It also became possible to introduce talks at the beginning of the academic year explaining the resources of the library to new students and, in fuller detail, what third-stage students should bear in mind when preparing dissertations. Both these pre-sessional routines were subsequently to appear in the programme of the Arts and Social Sciences Library.

After Jennifer Scherr left the Queen's Building staff to take on the role of Co-ordinator of Science Libraries in 1976, the remaining staff complement was enlarged to provide for one library assistant in charge of periodicals receipt and binding, and another to deal with cataloguing. A part-time library secretary also became available, dividing her work with similar duties in the Medical Library, the first such appointment to be made in any branch library and an indication of the level of business that the Faculty library now handled. A census taken in the 1975 session, showed 17,275 borrowings made by staff and students, 74,000 entrants to the library and 203 enquiries lodged at the enquiry desk, which resulted at times in the preparation of brief bibliographies and even some translation. Inter-library loans involved 785 volumes, a growing number of which were to local firms, especially the Filton Aircraft Companies. In turn, this led to developing contact between local technical librarians and Jeffrey Spittal, always free and informal during the 1970s and 1980s, but later put on a contractual basis and used to provide a source of income for the University.

In the early 1980s Marion Chandler joined the library, first as a cataloguer and soon after as Senior Library Assistant. A graduate in computer studies, she was in charge of its day-to-day running, whilst also liaising with departments over information needs, supporting users, and training students in information skills and the use of library resources.

Jeffrey Spittal retired in 1984 and Jennifer Scherr returned to the Faculty. She was now Sub-Librarian (Engineering and Science) in the

University, and responsible for a number of smaller branches serving the Science Faculty, as well as for engineering. Marion Chandler's role expanded to first Assistant Librarian and then Subject Librarian.

In the main reading area of the library changes were taking place. Computerised issues from the counter were possible from 1989 following the introduction of an automated library management system. Borrowers and library staff were at last freed from the chore of dealing with forms – some 20,000 loans then, but 55,000 by 2003! Inter-library loans could likewise now be requested electronically. The book stock was secured by an electromagnetically controlled and alarmed exit gate, and as a consequence, it was no longer necessary to continue the annual process of recalling stock from users and checking the shelves against the catalogue record.

During the 1960–90 period the Faculty had more than doubled in size. The new Merchant Venturers Building had been opened, there was an increased intake of post-graduates, and taught post-graduate courses had been introduced. All these changes resulted in greater daily working demands on the library, making it necessary to engage more library assistants, and even some students.

The accumulating demands were dealt with in 1998 by an increase of funding from Faculty sources, and this not only assisted book purchases, but also made possible the appointment of Patricia Coleman, as a second Subject Librarian; she took on responsibility for civil, electrical & electronic engineering, computer science, engineering management and engineering mathematics, with Marion Chandler as the senior partner liaising with the Faculty generally, and in charge of Aerospace and Mechanical Engineering, and the School of Mathematics in the Faculty of Science. To provide the necessary accommodation needed for their respective offices, and for a senior library assistant to act as branch supervisor, the librarian's office disappeared, along with two adjacent reading bays.

The advances in computer technology saw an increase in the availability of online library services, and in 2000 the University Library Services merged with the Computing Service to establish Information Services. This also saw a reorganisation of centralised management within library services, with greater communication and consistency between branch libraries.

In 2005 the library staff managed a major project involving the installation of rolling stacks in the Lower Reading Room, which provided a 50 per cent increase in shelf space, allowing the entire journal and report collection to be located there, in a subject sequence using the Library of Congress classification scheme. Classmarks for over 1,000 journal titles were created and added to the library's catalogue.

By the end of 2005 Marion Chandler had retired after 23 years in the Faculty library, and Meg Humphries, Assistant Subject Librarian, and Debra Avent-Gibson, Subject Librarian for Mathematics and Physics had joined the staff on a part-time basis. These posts were created to provide continuing subject help for readers in the Faculty, and in Mathematics. In the session 2005 the door counter registered over 100,000 and the loans and renewals totalled over 49,000, which meant that all three librarians were kept busy answering queries from the many users.

By 2007, staff and students were able to search out the contents of over 10,000 electronic journals from computer terminals in their rooms, as well as being able to carry out literature searches in indexes formerly only available in print, and to examine the holdings of many other academic libraries throughout the world. With the introduction of off-campus access to the University's network, the 'library' was no longer confined by the walls of the Queen's Building or to the 125,000 printed volumes in stock. The 'virtual library' is now a fact, and offers many more advantages and opportunities for cross-disciplinary research and teaching

More recent changes have been made to update the décor and lighting. The new lighting was partially funded by the University's Energy Office as the system is energy efficient. As daylight increases in the library, so the electric light dims' saving electricity. The library is currently seeking funding to refurbish its foyer area so as to create an informal study space, and its overall aim is to continue to meet the needs of the Faculty, and of the Department of Mathematics, particularly in implementing the Faculty's 2020 Vision of its future.

Chapter 10

One Hundred – Not Out, 1909–2009!

10.1 The Dearing Report

The report of the committee appointed by the Conservative Government under the chairmanship of Sir Ron Dearing (Ref. 15) was published in 1997 at a time of planned further reduction in the funding of higher education, which if carried through would have cut the unit of student funding by one-half since 1989. Fortunately, its implementation was the task of the New Labour Government elected in 1997, and it had very different ideas on the level to which universities should be funded. Of the many recommendations in Dearing's Report, the three which most concerned the university, were those dealing with the way in which it was run – 'Governance' it was called, the procedures used for recruiting students – 'Widening Participation', and the third was concerned with the use of the University's Intellectual Property (IP). Governance was not of immediate concern to the Faculty, and it is sufficient to note that action was taken in the University to expedite decision-making by a reduction in size of both Senate and Council. But the procedures to be used for recruiting high-quality students and the exploitation of IP certainly were of the utmost importance to the Faculty. In the first of these, the SARTOR (Appendix 2) requirement for its four-year MEng degrees – that at least 80 per cent of the intake must have obtained 24[1] points in their A-level examination – was a crucial

---

[1] For the period 1996–2006 the average UCAS score in the Faculty was 26.9, with 91 per cent satisfying the SARTOR requirement.

consideration. At this time, the University Plan envisaged an increase of student numbers to 13,000 by 2002, with a further increase to 15,000 by 2006, and it looked to the four-year degrees in Science and Engineering to play their part in this expansion. For the second of the important Dearing recommendations, the essential involvement by the Faculty with professional practice introduced what had usually been referred to as 'Outside Work', which itself required exploitation of the IP of academic staff.

Another recommendation of the Dearing Report (No. 34) also proved to be of extreme importance for the Faculty, so much so, that it is worth quoting in full:

> We recommend to the Government that it promotes and
> enables the establishment of a revolving fund of £400–500m,
> funded jointly by public and private research sponsors to
> support infrastructures in a number of top quality research
> departments which can demonstrate real need.

What followed from this recommendation is dealt with in section 3 of this chapter.

10.2 Widening Participation (WP)

10.2.1 The Faculty's Undergraduate Recruitment Policy, 1909–97

Reference to Chapter 4 of this history will show that the preferred requirement for entry to a Faculty degree programme until 1931 was successful 'matriculation', although acceptance was possible in these early years '. . . after passing some other test of fitness accepted in lieu thereof'. The great majority of students in the Faculty were either living in Bristol or in neighbouring counties, and would have taken the University's own School Certificate examination which satisfied matriculation needs. This situation – of a rather small, parochial, all-male Faculty, constrained in its ambitions by the cramped accommo-dation in the MVTC – continued until the Second World War. It is to be remembered that engineering was not commonly regarded as being a 'profession' at this time – to be counted alongside medicine and the law – and that the route to success was not necessarily a university degree, as it is now, but was also through an apprenticeship with an established firm, followed by progress through the examinations of the Engineering Institutions (Appendix 2).

The modest increase in student numbers of only 23 between 1911–31 (Table A4.1) indicates lack of pressure for admission to the Faculty, with no requirement save matriculation and ability to pay the fees. The large increase in numbers in 1946 compared to 1941 (Table A4.2) was due mainly to those returning from war service, for whom selection was a process observed more in the breach than in the observance, resulting in unusually large withdrawal and failure rates. But this was not so for the large increase between 1951 and 1956, which was attributable to the greater availability of financial support for university education provided by the Government through the Local Education Authorities.

It is necessary now to recall (Chap. 4) that within the three-year period 1943–6, Andrew Robertson had appointed four like-minded professors, who, when given the prospect of a new Faculty building, saw the opportunity to create in Bristol a Faculty of high ambition, a significant element of which was the recruitment of the best students on a *national*, rather than *local*, basis. Chapter 5.4 records the details of a 'Conference on Engineering Education as a University Study and as a Career', which they organised in 1955. It was an early exercise in 'widening participation', with a message to those whose school studies had not been in mathematics or the physical sciences, also emphasising that in the proper hands, engineering studies could be the basis for a broad education.

From the post-war period until 1964 it was necessary for the intending student to apply personally to each university, but by that year the increasing number of UK universities demanded a more efficient system, realised by the creation of UCCA – the Universities Central Council on Admissions – to which the schoolboy or girl applied, listing a maximum of five chosen universities. In 1993 the polytechnics were included, and it became UCAS – Universities Central Admissions Service, and the procedural details were also changed.

The competition for the best students produced a selection process in which the activity of assessing written applications received through the UCCA system, imposed additional loads on academic staff, which in the Faculty also involved a tour round the University, lunch with senior students, and an interview with the Admissions Tutor. It is easy to criticise the subjective components of such a process, but in choosing students the Faculty looked for personal achievement and potential, without reference to social or economic background. On this last issue, it was an established fact that engineering was a subject not often recommended to their best pupils by teachers in the independent schools because they themselves did not know what it entailed, and those that did appear from these schools were not often high academic achievers. But as the number of UK universities grew, an unofficial classification of them was reinforced, and it became politically fashionable to suppose that those at the upper end – Bristol being one[2] – were selecting students on grounds which favoured those educated at independent schools. The fact that the Government was beginning to consider requiring students – or their parents – to contribute to the full costs of their university education, was suggested as prompting some universities to act in this way.

10.2.2 The Dearing Recommendation (No.2) for Widening Participation and the Faculty's Response

Because the Dearing Committee had been established by Government, its recommendations carried serious implications for the future funding of universities, that concerning WP specifically so. Its second recommendation read:

> We recommend to the Government and the Funding bodies that, when allocating funds for the expansion of higher education, they give *priority* (my italics) to those institutions which can demonstrate a commitment to widening participation, and have in place a participation strategy, a mechanism for monitoring progress, and provision for review by the governing body of achievement.

From its early days in 1997, a component of the educational policy of the New Labour Government was that 50 per cent of young people in the 18–30 year-old cohort should have the benefit of a university education, and 'widening participation' was the phrase used. In pursuance of this ambition a substantial number of new universities were created from existing institutions of higher education, simply by a change of name. These were mostly colleges of further education, the

[2] Bristol is now a member of the 'Russell Group' of universities, whose mission is to conduct teaching within a research environment.

polytechnics having already been transformed into universities by the Conservative Government in 1993. In addition, although student maintenance grants were abolished in 1988 and replaced by loans, they were reintroduced in 2004 in order to help the WP initiative. These new means-tested maintenance awards were made available up to a maximum of £2700 p.a. where the parents' annual income was less than £18,360. Early in 2008 this last figure was increased to £25,000, with a cut-off at £35,000, and the maximum award increased to £2835 p.a., but later in the year the Government again found that its statistics were faulty and had to reduce these awards.

Whilst not forgetting that it had been early in the field through the establishment in 1970 of the 'University and Schools Committee' (Chap. 5.4), the University set up a group led by a Pro-Vice-Chancellor, to stimulate necessary activity in the general area of widening participation, using its share of the £10m which had been released by the HEFCE for this purpose. At the same time, the HEFCE had removed the MASN restriction on the number of students it would support at each university, whilst noting that it might be re-introduced if 'undesirable effects' appeared; presumably this anticipated the possibility that widening participation might be too successful!

What proved to be one of the most successful Faculty contributions to WP came through its participation in the University's annual Science, Engineering and Technology (SET) exhibition in the Galleries shopping centre in Bristol, where one of the several Faculty exhibits was the EERC stand showing the effects of earthquakes on buildings through a 'hands-on' model. Its ability to attract children (and their parents) was obvious, resulting in requests to take the demonstration into schools, and IDEERS – Introducing and Demonstrating Earthquake Engineering Research in Schools – was born. Its originators, Drs Wendy Daniell and Adam Crewe, held IDEERS 'build-and-test' competitions for many groups of schoolchildren[3] using the EERC shaking table, and received a £29,000 award from the EPSRC 'Partnership for Public Understanding' scheme. Since then, it has accepted invitations to take its activities to schools in Taiwan and Japan, with sponsorship from the Governments of both these countries. Two equally absorbing Faculty contributions to the SET exhibitions were provided by the Aerospace Department, one

---

[3]   And on two occasions for Members of the local Women's Institutes

being eye-catching air-flow movements around racing cars and sailing ships, and the other an urban light transport system, then at the initial concept stage, but which in 2008 was incorporated in the new Terminal 5 at Heathrow Airport.

Since the University started them in 1998, the Faculty, led by Dr Mike Barton, had taken a full part in the Summer Schools held in association with the Sutton Trust, founded in the mid-1990s by the philanthropist Peter Lampl, specifically aimed at encouraging able pupils at state schools to apply to departments in prestigious universities. In choosing participants for these Summer Schools the University gave priority to pupils who satisfied some of the following conditions:

(1)   they would be the first generation in their family attending university;

(2)   they had parents in non-professional occupations;

(3)   they attended a school which had a poor history of sending pupils to university during the last two years;

(4)   they attended a school which had a low overall A-level points score, and

(5)   they were plausible university applicants with a minimum of 5A*/A grades at GCSE.

Alongside the Sutton Trust there were, and still are, a number of other organisations having the objective of widening participation, amongst which is INSIGHT, run with financial support from local industry, aimed at tackling the perceived under-representation of women in engineering. By 2001 the Faculty had held its third INSIGHT one-week residential Summer School in which 40 girl sixth-formers visited the six Departments in turn, experiencing a mixture of lectures and practical work, with some visits to the companies which were supporting them. Fig. 10.1 shows a group of these girls holding their winning model land-yacht after it had been tested in the Aerospace wind-tunnel. Both the Royal Society and the Royal Academy of Engineering have their programmes for increasing the number of women students in science and engineering. In June 2005 the RS launched its Scientific Women's Academic Network (SWAN) with the University as a founder member, whilst the RAE has its own initiative for this purpose – described as HEADSTART – and there was also, until it was discontinued in 2007, a Government-funded National Academy for Gifted and Talented Youth

Fig 10.1 The INSIGHT Summer School. A group of girls holding their winning entry in the land-yacht competition.

(NAGTY), for which the Faculty hosted a Summer School in 2005 with the theme 'Mathematics in Modern Engineering'.

10.2.3 Widening Participation in the Faculty – Success or Failure?

The foregoing description of the Faculty's activities in WP indicates that it applied appreciable resources towards making it a success, intending to use as a measure any increase in the proportion of its students from state schools. The first recorded assessment was in 1999 when 60.7 per cent of its intake was from these schools, reducing to 56.7 per cent in 2000, but recovering again to 60.7 per cent in 2001. Within these Faculty figures there were distinct differences between Departments, with Aerospace and Computer Science accepting more state school pupils than the others. Table 10.1 gives a more recent assessment of Faculty success in this matter, and Table 10.2 gives the breakdown to Departments. Alongside these figures it should be recorded that the University's 'aspirational target' for state school students was 75.7 per cent, and that in the 2006 session it fell to 63.1 per cent from 65.1 per cent in the previous session. In both Tables 10.1

Table 10.1 Faculty results of the 'Widening Participation' activity, 2004–07

Session	Total Applications In all Categories	State School Applications	Total Intake	Intake from State Schools	% Intake from State Schools	Female Student Intake
2004	2929	2070	367	232	63.2	41
2005	2851	1935	381	239	62.7	47
2006	2794	1852	348	224	64.1	51
2007	3058	2121	365	231	63.3	54

Notes
1. The widening participation activities began in the 1998 session.
2. The second and fourth columns include all categories of student – private school, overseas, mature, etc., as well as state school.
3. The final column is for female students from all schools.

Table 10.2 Faculty intake numbers from state schools by department, 2005–07

	2005		2006		2007		Total	
Aerospace	49	(4)	43	(6)	45	(9)	132	(19)
Civil	32	(10)	41	(14)	49	(17)	118	(41)
Comp. Sci	65	(9)	47	(6)	46	(3)	153	(16)
E. and E.	28	(7)	28	(7)	21	(2)	72	(16)
Eng. Maths.	30	(9)	31	(12)	35	(18)	95	(39)
Mechanical	35	(8)	34	(8)	35	(5)	103	(21)
Total	239	(47)	224	(51)	231	(56)	673	(152)

Note
The entries in brackets are the number of female students entering from all schools.

and 10.2 the opportunity has been taken of giving information about the number of female students recruited from all schools during these four sessions; the significant increase does indicate some success for the many Faculty efforts to attract girls into Engineering, particularly in Civil, and in Engineering Mathematics.

One significant difficulty for the Faculty in achieving any WP target was the continuing insistence on high-level performance in the mathematics A-level. It knew from experience that although a greater proportion of 'A' grades were being awarded, the real level of achievement – relative to the Faculty's needs – was deteriorating, particularly in the state schools. Although denied by many of those involved, factual confirmation was given in 2008 by the Reform[4] think tank. It had made a study of GCSE papers in mathematics since 1970, concluding that '… it is now possible to achieve a grade C in GCSE maths having almost no conceptual knowledge in mathematics and by scoring less than 20 per cent in the top paper.'

An overall conclusion must be that the major expenditure of resources on WP, both at University and Faculty levels, has made little change to the proportion of students from state schools wishing to study engineering at Bristol, which can be taken to indicate that the Faculty's entrance procedures never did have any bias towards independent school pupils. The Dearing proposal for widening participation was an excellent examples of the 'Law of Unintended Consequences'. At a time when there was a national need for international-level excellence in university teaching and research in those institutions which were able to attract good students from whatever source, resources were forcibly used to increase significantly the numbers of central administrative personnel, whilst at the same time involving academic staff in duties which were not part of their prime functions, and which, in the out-turn, proved to be non-productive. It is likely that members of the Dearing Committee had forgotten any experiences which they may have had of university admissions procedures, and also that widening participation was not compatible with excellence in those universities which were fortunate in being able to attract the best students.

One unfortunate consequence of the University and Faculty efforts at WP was a perception generated by the actual efforts which were being made. They appear to have given an impression to the general public that a deliberate policy change had been made toward favouring pupils from state schools. The following quotation from *The Times* is indicative

> Bristol has long been a natural alternative to Oxbridge,
> favoured particularly by the independent schools, whose
> pupils take at least three in ten places. Departments are
> encouraged to make slightly lower offers to promising

---

[4] A prestigious independent, non-party, charity whose mission is to set out a better way to deliver public service and economic prosperity. The current Chairman of its Advisory Board is Sir Richard Sykes, Rector of Imperial College.

applicants with poor records at A-level, and some top schools have blamed the policy for the rejection of highly qualified applicants. As the most popular university in Britain in terms of applicants per place, however, Bristol has always had to turn away excellent candidates.

10.2.4   The National Situation

In its 2003 White Paper, 'The Future of Higher Education' (Ref. 16), the Government took a backward step on the WP issue by abandoning its 'aspirational targets' and rebuked the HEFCE for incorporating them into its strategy. With its continuing overall target of 50 per cent of the 18–30 year-old cohort into higher education, it said, '...we do not favour expansion on the single template of the traditional three year honours degree . . . but our emphasis will be on the expansion of two-year work-focussed foundation degrees.'

In 2007 the total Government allocation to the HEFCE for Higher Education was £7137m, and of this, £337m was allocated to WP, which, in turn dedicated £94m to widening access to full-time and part-time students, and £243m for *retention* (my italics) of these students.[5] What this actually meant in practical terms was not stated, but it is likely to have been for the extra teaching staff required to bring the weak students up to an acceptable level. Evidence in the Faculty on this issue, relating to mathematics, indicated great difficulty in building on the deficiencies reflected in poor A-level results. In explanation of the method for apportioning the £243m to the 130 institutions which it funded, the HEFCE assigned first-year students to one of six categories based on age and qualifications at entry. At the 'high risk' end of the retention scale were mature students with 'non-traditional' entry qualifications or low-grade A-levels, and at the 'low risk' end were the type of student recruited into the University and the Faculty. Despite these attempts at social engineering, which had resulted at this time in 43 per cent of the 18–30 year-old national cohort entering higher education – and the expenditure of around £800m for WP over a five-year period – retention levels had not improved. In February 2008 the House of

---

[5] For these two activities, in 2006 the University received £214k and £487k, respectively. It also received £64k to provide access to students with disabilities. Although the total HEFCE award in 2006 showed an increase of 7.1 per cent over the previous session, for WP it received 12 per cent less.

Commons Committee of Public Accounts (PAC) reported that 'Since the Committee last reported in 2002 there has been no reduction in the percentage of students in England not completing their higher education course *at their original institution* (italics are mine): the figure remains at 22%.' In presenting these figures the cross-party PAC blamed the high non-completion rate on the drive to widen participation, with the fee-structure tempting universities to admit poorly qualified students. The percentage given was for full-time students; for those taking a degree on a part-time basis, the figure was 45 per cent. The phrase in italics in the above quotation should be noted, because students who transferred to another university – and were eventually successful – would reduce the figures given by the PAC by non-negligible amounts (Table 10.3). Looking specifically at withdrawals at the end of the first year, the PAC commented that the Russell Group of universities had high continuation rates, whereas at the other end of the scale there were 12 institutions which had rates below 87 per cent.

It is likely that the PAC used data provided by the Higher Education Statistical Agency (HESA), and it is useful here to reproduce their information for non-continuation following year of entry, the year chosen being 2005; it has the advantage of differentiating between 'young students' and 'mature students', whilst also dealing with the 'transfer' issue. Table 10.3 summarises the situation for the UK, England, and our own University, giving numerical values both without, and with, the 'transfer' effect. Remembering that these figures are only for non-participation after the first year of study, they also point to the distinct possibility of the PAC's 22 per cent figure being achieved during a study period which might extend over five, or more, years.

Table 10.3 From HESA Tables; Non-continuation following year of entry. Full-time first degree entrants 2005 session.

	Young Entrants		*Mature Entrants*	
UK	9.8	7.1	11.3	8.7
England	9.5	6.7	11.0	8.3
Bristol	4.4	2.8	4.9	3.4

Note
The first figure in each main column is for comparison with PAC data, i.e. transfers to another HEI count as non-continuation; whereas the second figure in each main column accepts such transfers in assessing non-continuation, so that they are always smaller than those in the first column.

Obtaining reliable non-continuation figures for the Faculty presents considerable difficulties. In addition to the discovery by some students that engineering is too demanding a field of study – resulting usually in transfers to other faculties in the University – there are well-established routes for first-year transfers between Faculty Departments, transfers between three-year and four-year courses, and the intercalation of year(s) in industry for some courses. The most complete, and easily understood, data available, is that obtained by Dr Martin Lings, Director of Studies, over a ten-year period for Civil Engineering (Fig. 10.2). Here, he has followed the detailed progress of each student during their time in the Department, having entered in one of the sessions 1996 to 2004 as indicated. The triangles and crosses signify percentages of the three- and four-year students, respectively, who have been successful in obtaining their degree, whilst the full circles represent the percentage of students in both groups who left the department; thus, for each session, the sum of ordinates is 100 per cent. The conclusion here is, therefore, that the rate of failure to obtain a degree in Civil Engineering in the Faculty varies yearly between 5 and 15 per cent over the period studied, but averages around ten per cent. Some data is also available for Mechanical and Electrical Engineering in the Faculty. For

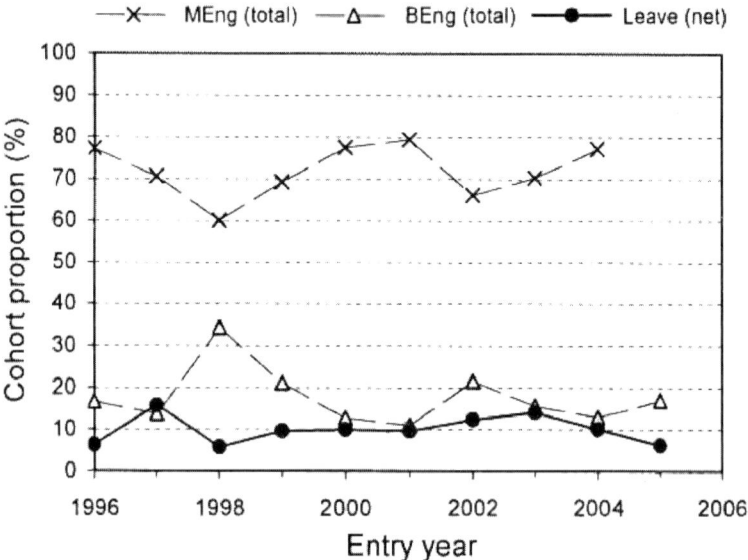

Fig. 10.2 The ten-year study by Dr. Martin Lings of progress by Civil Engineering students.

the period 1995–02, 'failure' on the four-year MEng course in the former averaged 16.3 per cent, but when transfers by the 'failed' students to the three-year BEng course is included, the failure rate drops to 4 per cent. It is necessary to record here (Table A.13) that in this Department there is no stated preference at entry for the length of course; the decision is made on the basis of first-year performance. For Electrical Engineering, the average non-completion rate for the same period was 8 per cent. When it is remembered that 'transfers-out' of Engineering to other Departments in the University are historically considerably larger than 'transfers-in' – which adds to the non-completion percentage – these figures, which are representative of the whole Faculty, are not inconsistent with the University figures given in Table 10.3.

In the broad spectrum of English higher education institutions the effect of attempts at WP can be seen from the way in which the HEFCE distributed its funds for the 2008 session. The Government had awarded it a 3.3 per cent increase over the previous session, against an inflation rate of 2.75 per cent, but the majority of those universities which found themselves in the various lists of 'top universities' were lucky if they received sufficient to cover inflation (Bristol received 2.8 per cent). But those that had the highest proportion of students from the lower socio-economic groups – and were generally low down in the lists – were well treated, often receiving double-figure increases.

10.2.5 The Engineering Diploma

A major government initiative having widening participation as its aim, was the introduction in 2005 of Diplomas for 14–18 year-old school-children, as an alternative to the GCSE and A-levels, in 14 different subject areas. To quote the government's own words

> Diplomas will give young people a real alternative to
> traditional learning styles by offering an imaginative, high-
> quality blend of general education and applied learning.
> Diplomas will take young people where they want to go, either
> to further or higher education or to the world of work, with
> equal esteem.

Compulsory elements for all diplomas include 'functional skills in mathematics, English and IT, with a minimum of ten days' work experience to introduce teamwork and self-management skills.' Two of the first five of these diplomas, to be started in 2008, are in Construction and in Engineering, at three levels which are said to be equivalent to 4–5 GCSE, to 5–6 GCSE, and to 3 A-levels, respectively. In the Engineering Diploma, although it is stated that Level 3 will 'provide learners with an academically rigorous programme of study sufficient to recognise progress into higher education', its mathematics content is considered to be sufficient only for Technician Engineers (Appendix 2), and perhaps for the two-year 'foundation degrees' envisaged in the 2003 White Paper (Ref. 16). In the light of this, the Royal Academy of Engineering (RAE) were quick to point out that such a mathematics syllabus was an inadequate preparation for the university degrees which led towards professional status, and accordingly has produced its own more advanced syllabus for the Diploma pupil to take if they wished to apply to study engineering at university.

For the provision of the ten-day work experience, it seems that the collaboration of industry was taken for granted by the Government, and initially the CBI supported the concept in principle. But it subsequently made known its considered view, that efforts ought to have been made to consolidate the value of the GCSE and A-levels, instead of introducing something completely new, bringing with it the danger of introducing a fractured system where the independent schools retained A-levels, whilst the state schools were forced by Government to opt for the Diploma.

The initial response of the Faculty to the engineering diploma, is that it will be considered for admission, subject to an A-grade overall, and so long as it includes the RAE special mathematics unit, plus a combination of one A-level and one AS-level in subjects appropriate to the Department concerned. An exception to this general rule is that for entry to the Engineering Mathematics Department, an A-grade in A-level Mathematics will be required. Clearly, the Government's introduction of the Diplomas will do nothing to simplify the task of schoolteachers in advising their pupils on subjects to study with university entrance in mind, nor for the pupils in making their choices!

10.3 The Bristol Laboratory for Advanced Dynamics Engineering (BLADE)

During the 1990s it had become clear to all those concerned with the excellence of UK science and engineering education at university level, that the large number of universities which had recently been created, had, in equipping their new laboratories, absorbed a disproportionate share of the finance available. As a consequence, similar laboratories in many of those universities which had been long established and were regarded as being of the first rank, were working with out-of-date equipment and facilities. To remedy this situation, the JREI initiative referred to earlier, was created; but, of even greater importance, a much larger fund of £600m was provided jointly in 1998 by Government and the Wellcome Trust, to be bid for competitively in a laboratory refurbishment scheme organised by the HEFCE. The existence of this fund can be traced back to Recommendation 34 of the Dearing Report, as has previously been mentioned in the first section of this chapter. The awards, from what became known as the Joint Infrastructure Fund (JIF) were to be made to enable world class UK engineers and scientists to remain at the leading edge of research. The initial UK allocation to engineering and materials science was £59m.

If one refers back to the Faculty's early beginnings in Unity Street, the very acute shortage of space there, coupled with the small size of the individual Departments, meant that research space and workshops for producing test-pieces were common facilities. In the move to Queen's Building a large central workshop – run by Mechanical Engineering but in principle available to all Departments – was supplemented by individual workshops, each with its own staff of technicians. As Departments grew, so did the isolation of the different workshops, and when electronic instrumentation replaced mechanical systems, each Department created its own unit to cover this need.

Between 1970 and 1990, several Deans attempted to recreate a Faculty laboratory and workshops structure, but all failed, partly because of the sense of belonging to a particular Department which the academic and technical staff shared, but also due to lack of funding which the intended rationalisation would have required.

The above situation is recorded here because one of the conditions set down for JIF bids was that they should be based upon the common features of an engineering research programme across a broad spectrum, rather than those of a special sector. In the Faculty this was not difficult, because all Departments had long-standing interests in the behaviour of engineering systems under dynamic loads. Both Profs. Pugsley and Collar had earned enviable reputations in the dynamic behaviour of aircraft, and the former had extended this research into the area of flexible bridges. His successor had established the Faculty as the prime UK centre in the field of experimental structural dynamics, and in Aerospace the Department had developed much sought-after facilities in industrial aerodynamics, in helicopter dynamics and space vehicles. In Mechanical Engineering, the studies by Morrison et al. on the fatigue behaviour of metals under repeated loads at high pressure had been followed more recently by a research centre devoted to real-time control engineering, which was even then being used in collaboration with other Departments in the Faculty. In Engineering Mathematics new developments had been made in the mathematical aspects of non-linear dynamics and in the theory of chaos, and at Faculty level all these research advances had been made possible by interaction with colleagues in the Electrical Department.

There was, therefore, contributions from all parts of the Faculty in the general field of structural dynamics, but a specially important role was that played by the Earthquake Engineering Research Centre (EERC). Not only had it become a focus for earthquake engineering research in the UK, but it had also been invited by the European Commission to coordinate its programmes of experimental research in this field over a period of ten years, and in fulfilling this role it had collaborated with all the major European laboratories in basic research, and in immediately applicable studies for industry. But it was collaboration actually within the Faculty which placed the EERC in a pole position, most notably with the control engineering group led by Prof. D. Stoten, whose basic research allowed the EERC shaking table to be operated with greatly increased precision, and in 'real-time'. This achievement, at the cutting edge of Mechanical Engineering at world level, took the EERC into collaboration with US research centres, leading to invitations from the US National Science Foundation to take part in its own $80m NEES (National Earthquake Engineering Simulation) programme. It also allowed Prof. Stoten to accept invitations from Japan to participate in the development of the world's largest (1000T) shaking table at Miki City.

With this background of proven national, European and internationally successful collaboration, it was realised that a suitable theme for the JIF bid would be 'Response of Engineering Systems to Dynamic Loading', and that future success would depend on establishing the coherence of laboratory facilities which had previously been lacking.

The actual achievement of a programme to achieve this would rely on the willingness of the laboratory staff to assist in a complete restructuring of the Faculty's laboratories, and the loss of their affiliation to a particular department. On the other hand, the funding to be made available would allow them to equip the workshops and laboratories with the most modern equipment.

The Faculty's successful JIF bid was awarded £15m, more than one-quarter of the total available nationally for engineering, to which the University agreed to add around £5m from its own building funds. In essence, the building programme consisted of the creation of three large, open-plan laboratories, one in the basement of Queen's Building for control engineering, and one in each of the two original car parks – one for earthquake engineering, and one for aircraft structures. For these last two, due to the nature of the research to be carried out, deep excavations into the foundations was required in order to provide the necessary headroom for large testpieces. For the earthquake laboratory this meant that the lowest three floors, of a five-storey building, were constructed below ground level, giving it the unique feature of exceptionally strong walls which can be used as reaction-walls for testing purposes. In Fig. 5.3 attention has already been drawn to the two car parks belonging to Queen's Building; Fig.10.3 is an aerial view (courtesy of Google Earth) showing the two BLADE laboratories which were constructed on these car parks. It also shows, at top-centre, a view of the Merchant Venturers Building (Chap. 9.4)

In addition to the three new research laboratories, the teaching laboratories were completely transformed for Faculty, rather than departmental, use, in almost all cases with new equipment. But perhaps the most significant change was in the re-organisation of the laboratory and workshops support staff; now, they were all part of a faculty system, organised under a single management structure (Fig. 10.4) with three main groups – test laboratories, materials and workshops, each one having the most modern machines and other facilities for their work. It must be recorded that the technical staff themselves initiated this change, and were responsible for carrying it out during an 18-month period in which teaching and research continued amidst organised chaos.

The foundation stone for the new laboratories, which together now constitute the Bristol Laboratory for Advanced Dynamics Engineering (BLADE), was laid by the University Chancellor, Sir Jeremy Morse, and it was officially opened on 25 February 2005 by HM Queen Elizabeth II

Fig. 10.3 Queen's Building, showing two of the BLADE laboratories built on the car-parks; the third, linking these two, is in the basement below the entrance hall. The new Merchant Venturers Building can be seen at top-centre (Courtesy of Google Earth).

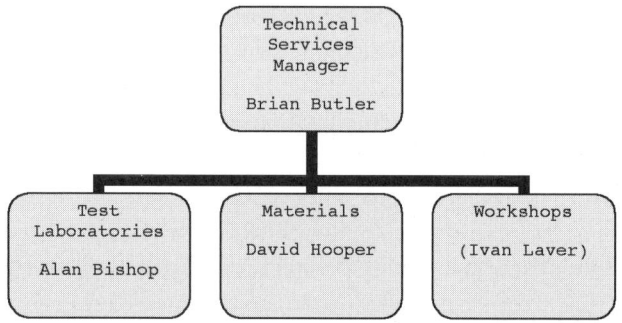

Fig. 10.4 The BLADE re-organisation of workshops and laboratories.

(Fig. 10.5), a little over 46 years since she had opened Queen's Building itself. In Fig. 10.6, Her Majesty is shown being presented to the three Faculty Deans (Muir-Wood, Severn and McGeehan) who were involved in different stages of the project.

Fig. 10.5 The BLADE Project. HM the Queen unveiling the stone at the official Opening.

Fig. 10.6 The BLADE Project. HM the Queen being introduced by the Chancellor Baroness Brenda Hale to three Faculty Deans, David Muir-Wood, Roy Severn and Joe McGeehan, who had been involved in the project.

In such a large and complex laboratory it was inevitable that there would be difficulties in implementation, and so it proved with the hydraulic system common to all facilities. One effect of this was that the MSc course in 'Advanced Dynamics Engineering' – approved by Senate in 2004, has not yet been offered, but is likely to be so now that Prof. David Ewins has been appointed as the overall director of BLADE.

At the conclusion of the JIF opportunities, the HEFCE continued with its aims of upgrading research facilities on a competitive basis by the introduction in 2001 of its Science Research Investment Fund (SRIF), which has been repeated at two-yearly intervals, and which initially required universities to contribute 25 per cent of the total from their own funds, but was later reduced to ten per cent. In 2003 the University attracted £25.97m from the second offer of this fund, two projects coming to the Faculty. These were £3m to Prof. McGeehan for 'Office and Research Space for the Optics, Radio Frequency and Image Processing groups in the Merchant Venturers and Queen's Building,' and £107,000 to Dr Andrew Hogg for 'Equipment required for research in Thermodynamics and Fluid Mechanics' which could not be provided in the BLADE project.

In 2008 the HEFCE discontinued its SRIF fund, replacing it with two new capital investment streams, one for 'Teaching and Learning', and one for 'Research'. Between 2008–11 the University expects to receive £53m from these two streams.

10.4 Postgraduate Education for Industry

10.4.1 Introduction

The first mention of postgraduate education in the Faculty occurs in its prospectus for 1937, when notice was given that 'The course, extending over one session, will cover a comprehensive treatment of the mathematical and mechanical methods of analysing the stresses in redundant structures, including framed structures.' The notice went on to say '... that prominence will be given to the work of the Steel Structures Research Committee, attention being paid to the effect of recent advances in methods of construction and in continental practice.' Both Andrew Robertson and John Baker (later Lord Baker of Windrush), then Head of Civil Engineering, were members of that Committee, during which time Baker was beginning to formulate his ideas on new methods of design which utilised the plastic deformation of steel, ideas which through his design of the Morrison[6] air-raid shelter, were to save many lives in England during German bombing raids in the Second World War. These popular courses, being complemented by laboratory tests on full-scale joints and model tests on complete structures, were repeated annually until war intervened in 1940. But there was no mention of the possibility of degrees being awarded. Even though the MSc existed in the Faculty, its regulations made it clear that it was a degree associated with research, rather than with a taught course.

During the protracted post-war period, the Faculty was busily occupied with the building and occupation of Queen's Building, but in 1959 it did express an interest in the creation by the University of a Board of Management Studies, to be funded by industry after an initial pump priming. John Morrison was the Faculty representative on this Board, where it soon became clear that the dominant interest was towards the social sciences end of the academic spectrum – a fact which may have contributed to its short life.

In 1964, the first Personal Chair in the Faculty was awarded to Gordon Rogers as Professor of Thermodynamics in Mechanical Engineering. One of his many qualities was that of real excellence as a teacher, evidenced by his book *Engineering Thermodynamics* (with Jon Mayhew) which became the standard text in many countries, and it was therefore natural that the UK nuclear power industry should turn to him to provide postgraduate education for its employees. As a result, the Faculty's very first postgraduate course was created in 1965, with the title 'Fluid and Thermal Studies for Industry', and until Rogers' retirement in 1982 the Faculty had a small nuclear engineering laboratory. Soon after, following the run-down of the UK nuclear industry and the consequent reduction in staff in the thermodynamics area, the MSc was given a broader sweep with the title 'Advanced Mechanical Engineering' allowing contributions from what were now its three main areas of expertise

---

[6] Most politicians would like to be remembered by eponyms, and here are two. Herbert Morrison was Home Secretary in Churchill's wartime Coalition Cabinet. John Baker's shelter was a steel dining-room table under which the family could shelter during an air-raid. It proved capable of taking a collapsed house without itself collapsing. After the war, Baker was awarded £3000 by the Commission for Awards to Inventors. Sir John Anderson, who was Home Secretary and Minister of Home Security in 1939, was credited with the 'Anderson' shelter, even though it was designed by a fellow Scot, Sir William Paterson. This pre-fabricated air-raid shelter – three million of which were made – consisted of curved corrugated-iron sheets which were bolted together by the home-owner and partly buried in the garden, with soil and turf as cover.

– dynamics and control, design and process engineering, and solid mechanics. Other Departments now began to see advantages in offering MSc courses, with Computer Science offering four, Electronics and Civil one each; the Aerospace IGDS MSc will be discussed shortly. Civil had also obtained approval in 2002 for an MSc in 'Civil Engineering and Construction Management'; it ran for only one year, during which time it was rediscovered that the project work and dissertations were too demanding on academic staff time for the little financial reward which resulted; it was formally withdrawn in 2008.

10.4.2 The Rise and Fall of Management Studies

The role of the Engineering Council in 1982 in setting down the requirements for the further education of graduates who wished to become Chartered Engineers is discussed in Appendix 2. One topic on which it placed great stress, that of the management of engineering enterprises, was taken up by the IMechE which instigated for its members a distance learning qualification described as the Diploma in Engineering Management. It was earned by satisfactory performance in modular courses which the IMechE had persuaded a number of universities – including Bristol – to offer. Within a short time, the other major PEI had joined the IMechE to establish the Joint Board of Engineering Management (JBEM).

In July 1994 the JBEM visited a number of universities, having in mind the establishment of an MSc in Engineering Management, as a result of which Bristol and Manchester were selected. The Faculty's MSc in this subject was begun in 1995 on the basis of two years of part-time study plus a dissertation, with entry made conditional by the JBEM on the possession of a Postgraduate Certificate in Engineering Management which included such topics as human and financial resources, and project and marketing management. A temporary disadvantage for the Faculty in promoting this MSc was that its Ordinances and Regulations did not include postgraduate *certificates* (my italics) of any kind, but by the 1999 session this drawback had been removed, and, in case it should be needed in the future, a postgraduate diploma was approved at the same time.

Most of the teaching staff offering the Engineering Management MSc joined the Faculty from other Departments in the University, or were engaged on a contract basis to provide the professional studies courses, and together were described as the 'Engineering Management

Group'. With early success, the ambition emerged to extend the scope of the MSc towards the EngD – a new qualification which had been recognised by the Quality Assurance Agency, and which will be discussed in detail shortly. To do so required the agreement of the Engineering Institutions to allow the JBEM to be replaced by an Engineering Management Partnership (EMP), made up of Bristol and the universities of Loughborough, Leeds and Edinburgh, the first of these already having an EPSRC-sponsored EngD in 'Innovative Construction'. From 2001, therefore, the Faculty offered postgraduate taught programmes leading to the Certificate, the MSc or the EngD, depending on the enthusiasm and competence of the candidate, for whom the last of these degrees involved a three-year commitment full-time, or six-year part-time. It was also in 2001 that the group were authorised to offer an MSc in 'Entrepreneurship' in conjunction with the Bristol Enterprise Centre (see section 10.5.3 of this chapter).

But these ventures were not a success. Between October 2001 and March 2006 only 19 students had been recruited, and several of the key staff had become unavailable and could not be replaced due to lack of income. Recruitment was suspended in 2006 and in August 2007 the remaining staff were transferred to Civil Engineering, with both the MSc and EngD courses being formally discontinued in May 2008.

Before leaving Management Studies, it is noted that in May 1991, the then Registrar, Cathy Cunningham, proposed a 'Graduate School of International Business' in collaboration with the *Ecole des Ponts et Chaussées* in Paris, but without any contribution from the Faculty. Its aim was the MBA degree, but it failed to excite any general interest in the University and so did not succeed. Again, in October 2000 the Senate discussed creation of a 'Management Research Centre', to be located in Social Sciences in the first instance, but having strong links – it was said – with other faculties, particularly Engineering, but these did not materialise.

10.4.3 Research Council Support for In-Career Postgraduate Education.

In addition to their role in providing grants for research, the Research Councils have always supported schemes of postgraduate education and training, usually aimed at some form of university-industry collaboration. An early arrangement of this kind was the Teaching Company Scheme (TCS) involving a one-to-one relationship – a negotiated agreement in fact – between an HEI and a company. Its principal activity was the solving of a particular problem being experienced in that company, by secondment of one, or more, of its graduate employees on a part-time basis to an HEI which had proven research expertise in that problem. The contract was normally for two years, during which time the employees would divide their time between HEI and company, as the progress of the project required. The cost of the TCS was borne approximately equally between the SERC and the company, and its management required formal six-monthly meetings, with progress reported to the SERC. When successful, the TCS effected transfer of cutting-edge research from the HEI to the company through the learning process of the graduate employees. The nature of the TCS did not normally allow them to obtain a PhD, but in most cases an MSc by research would be awarded. One requirement was open publication of results, so that no question could arise of public funds being used for private gain.

The Faculty was involved in five TCS. In Mechanical Engineering in 1984, the first two TCS in the University were with Avon Rubber plc; one to investigate 'The modelling of the rheological properties of extruded rubber', and the second to implement 'Modern concepts in automatic control to mixing, extrusion and other manufacturing processes'. The Aerospace TCS with Westland Helicopters was for 'Developments in thermoplastics technology', whilst in Civil Engineering the Earthquake Group linked with the Ove Arup Partnership for the 'Experimental validation of seismic design software', which had been written by Arup to the most up-to-date codes, but had not received any actual verification. The fifth Faculty TCS was between the Grinding Institute (a unit within Mechanical Engineering) and Engineering Mathematics, with RHP Aerospace of Cheltenham, on 'Development of an expert system to model the grinding process'.

The Cooperative Awards in Science and Engineering (CASE) and the Integrated Graduate Development Scheme (IGDS) are also programmes which aim towards satisfying the needs of industry for an advanced level of skills, rather than the transfer of research results as in the TCS. The original CASE bore some similarity to the TCS, but with a shorter period spent in industry, and financial support from the Research Council rather than a salary from industry. The current awards of this nature, described as 'Industrial Case', are for a three-and-a-half year period, only three months of which, as a minimum, are to be spent with the sponsoring company. This last requirement is presumably to recognise the fact that some university regulations for the PhD degree require three years in residence, though this is not so in Bristol.

The IGDS is a professional development scheme directed towards graduates in a particular industry who already have substantial industrial experience; and to emphasise this, the participants are described as 'delegates', rather than students. The course, made up of modules, leading to the MSc degree, can be spread over a five-year period. In the Faculty, the Aerospace Department, jointly with the University of the West of England, acts in partnership with all the major UK companies in aerospace engineering in offering the only IGDS in this industry. In 1997 it received an award of €139,000 from the European Commission to give its IGDS a European dimension with the acronym EUROPADS – European Professional Aerospace Development Scheme, having members in Belgium, Germany, Ireland, Italy and the Czech Republic, but this ambitious scheme proved to be unworkable, largely due to the underestimated task of dealing with different languages and administrative practices. In 1999 a further development was Senate's approval of the accreditation of Rolls-Royce in-house courses as part of the IGDS, and from May 2005 the UK activity was broadened to allow non-graduate engineers to participate if they had sufficient years of industrial experience following NVQ, HNC or HND qualifications. This arrangement was then referred to as the 'Bristol and West of England Consortium for Continuing Professional Development (CPDA)'. Because a number of overseas students preferred spending four years of study in obtaining two degrees rather than one – a BEng plus an MSc, rather than an MEng – in 1995 the IGDS modules, plus a project and dissertation, were formed into a full-time second-degree course. A final development of the IGDS concept was provided by the avionics modules; these – a mixture of aeronautics, electronics and computer science – became so popular that they formed the basis of new degree courses at both BEng and MEng levels.

In 1998, as part of the University's development plan, a Faculty initiative was accepted by the EPSRC to develop an IGDS in 'Competitive Product Engineering' for the SME in the south-west of England involving ten other HEI in the region. Unfortunately, once again, the administrative cost of coordinating the large number of possible partners proved too great, and the scheme was never started;

unlike the large companies, the SME did not have the resources to make an effective contribution.

10.4.4   The EPSRC Engineering Doctorate (EngD)

The EPSRC describes the EngD as its 'flagship doctoral qualification'. It was introduced in 1992 following the Report of a Committee[7] chaired by Prof. John Parnaby, with the aim of providing high-level knowledge transfer through people in academe and industry. In fact, the relevant section of the report reads 'A doctoral programme in Engineering alongside the traditional PhD, containing coursework, research and a project designed to give the candidate experience in teamwork and engineering project management.' The EngD is an extension of the Teaching Company Scheme described above, because in specifying its details the EPSRC states 'The work has to be a significant contribution to the work of the Company'. One difference, however, is that instead of a one-to-one link between HEI and company on a specific topic, the EngD offers a combination of HEI and companies the possibility of dealing with a generic theme over a longer period of time. In so doing, it does of course, as compared with the TCS scheme, increase the management effort required to set it up in the first place, and then to organise on an individual basis the recruitment of the young engineers taking part, coupled with the interface between them and the company involved. A second difference is that the researcher in the EngD scheme is not necessarily already an employee of one of the sponsoring companies, as was the case for the TCS, but may have been recruited through an advertisement and then 'sponsored' by a participating company.

By its title, the EngD is intended to be more prestigious than the MSc, and more indicative of its purpose than the long-established PhD degree. In furtherance of this intention, its participants, who are graduates with a high quality first degree, are described, not as postgraduate students, but as 'Research Engineers (RE)', and their financial support comes not as a salary or grant, but as a stipend.

An EngD programme lasts for four years, and is carried out with a university as a focus, interacting with a number of industrial partners, each supporting one or more RE, so that one of its novel features is participation by a group of RE carrying out research projects on a generic theme. Whereas the industry partners are responsible for producing

---

[7]   The author was a member of this committee.

the research topics, which lie in the direct line of their activities, the university provides a series of modular taught courses related to the generic themes, and also provides a physical centre where RE and their sponsors can meet with university staff. Such a common meeting ground is an important feature of the EngD scheme because one of its requirements is that the RE must spend at least 75 per cent of the time carrying out research, not in the university, but in the premises of the sponsoring company.

As with all its user-focussed training activities, the EPSRC has, since 2003, funded EngD schemes through Collaborative Training Awards (CTA) which it makes to universities and other research organisations. These provide flexibility over a four-year period for planning and for responding more quickly to changes in postgraduate training opportunities without the need for detailed approval by the EPSRC.

By 2007, a total of 22 EngD centres had been created in the UK based on 14 universities. They had enrolled 1230 RE, sponsored by 510 different companies, 28 of which had supported more than one RE.

10.4.5 The Bristol/Bath EngD Centre in Systems Engineering

A convenient definition of a 'system' is that it is an integration of elements from which new behaviour emerges. In engineering products the elements are materials of different kinds, and so the system might be described as 'hard'; but its performance is usually influenced by the qualities of the people who design, make and manage the product, which allows their contribution to be described as 'soft'. The total system therefore consists of both 'hard' and 'soft' elements. In engineering education at undergraduate level, most of the formal courses deal with the 'hard' aspects, whilst the 'soft' aspects can derive from all the other benefits of life at a university if the student takes the opportunity to do so. Both aspects are contained in what the Engineering Council describes as 'the formation' of the engineer, progress in which has been vested in the requirement of Continuing Professional Development (CPD), the middle word here signifying an enhanced proficiency in knowledge of both 'hard' and 'soft' systems.

The Bristol/Bath EngD Centre has its origins in the Systems Research Group created in the Faculty by David Blockley, and aims towards the CPD needs of young engineers of the highest quality who have the potential to be industrial innovators and managers of the future. In very broad terms, through the integrated approach of the BLADE initiative, the Faculty will deal with the 'hard' systems, whilst the Bath Business School will cover 'soft' systems.

The £3.4m award to the EngD Systems Centre was made in December 2005, allowing the first cohort of 13 RE to start in January 2007, the intervening period being spent on discussions with 58 national, European and international companies, and in forming a Strategic Advisory Board as required by the EPSRC, with Patrick Godfrey – now a part-time Professor in the Faculty, but previously with the Halcrow Partnership – as its Director. In its membership are senior representatives from twelve leading UK companies, six from the two universities, and one each from the EPSRC and the SW Regional Development Assembly.

In the first cohort of 13 RE, five were employed by industry, which meant that they continued as such, and only their fees were a charge on the £3.4m award. The remaining eight were recent university graduates with either the MEng or MSc degree; these RE are described as being 'industry-sponsored', requiring that both fees and stipends be paid from the award. In the 2006 session the stipend was £13,800 p.a., plus an additional mandatory contribution of not less than £3000 p.a. from the supporting company. For all practical purposes, including the ownership of the research undertaken, the industry-supported RE is regarded as a company employee.

During the four years of the award, the overly ambitious aim, bearing in mind the need for individual arrangements, was for an annual intake of 25 RE, and since the funding allows for only ten industry-sponsored RE, the remaining 15 would have to be industry-employed. The second cohort – for the 2007 session – reflected more realistic ambitions and consisted of 12 RE, three of whom were 'industry employed' and nine 'industry sponsored'. For the 2008 session the recruitment situation had clearly begun to reach an achievable level and ten new RE were recruited, three of whom were 'industry-employed'.

10.4.6 Other Faculty Involvement in EngD Schemes

(a) Non-Destructive Engineering (NDE): In 2001 an EPSRC Research Centre in NDE was formed from the six universities at Bath, Bristol, Imperial College, Nottingham, Strathclyde and Warwick, collaborating with 12 industrial organisations. In 2004 the same group was invited by the EPSRC to make a successful bid for an EngD Centre. In 2008, the

Faculty had four RE in this centre, sponsored by Airbus, Rolls Royce and Quinetiq, working in ultrasonic arrays, acoustic emission, structural health monitoring and non-linear NDE methods, with corresponding taught elements in these areas.

(b) Nuclear Engineering: Again, at EPSRC's request, in 2006 the universities of Manchester, Imperial College, Bristol, Leeds, Sheffield and Strathclyde were asked to create an EngD programme in nuclear engineering. The first two of these universities were to provide the taught course components, with the remainder providing the research expertise. Thus far, despite the national need for skills in this area, the Faculty has not been able to recruit any RE for this EngD.

(c) Aerospace: In a previous section of the chapter (10.4.3) it was recorded that the very successful IGDS – referred to as CPDA – had been developed to provide an MEng degree course, and it is this which has been used as a basis on which to provide an EngD on 'Aerospace Design Manufacture and Management' with UWE and the Filton group of aerospace companies.

In concluding this section on the EPSRC EngD it is recorded that its potential for further contributions to research and postgraduate education benefitting UK industry was appreciably extended in December 2008 by a £250m Government initiative for creating 44 centres which, in total, would provide 2000 PhD and EngD graduates in areas of national importance such as climate change, energy, an ageing population, and high-technology crime. The University received four of these awards – only University College London receiving more – two in the Science Faculty and two in Engineering. These last two were in areas where EPSRC-supported programmes had already been started, the EngD in Systems in Civil Engineering, and the Advanced Composites Centre for Innovation and Science (ACCIS) in Aerospace Engineering, each receiving around £6m for the extension of its activities.

10.5 Outside Work: Exploitation of Intellectual Property

10.5.1 Introduction

The following quotation relating to outside work is to be found in the Dean's annual report for 1911:

> While the extension of knowledge by means of research work is
> an important function of any Faculty of a University, it is
> specially incumbent on an Engineering Faculty to be of service to
> industry in general, and more particularly to those engaged in
> undertakings in the district in which the University is situated.

The Dean was of course Prof. Wertheimer who was also Principal of the
MVTC, and was, in the quotation, presenting his intention of continuing strong links with industry, which may have caused concern in some
of his new colleagues from the former UCB. In illustration of his views,
he goes on to list work in the form of tests on materials which had been
undertaken by the Faculty, together with experiments on engines for
motor cars, tractors and aeroplanes carried out by Prof. Morgan, Head
of the Department of Motor Car Engineering, and development of an
electric impulse clock by Prof. D. Robertson, which had been marketed
by Messrs Grant and Co. of Leicester. It is unfortunate for our present
purpose that no details are given by the Dean of financial, or other
arrangements through which the work was carried out; were the academic staff paid in any form, or did they undertake to do such work
as part of their terms of employment? We do know that local industry
gave laboratory equipment to the Faculty, because such gifts were
recorded in the annual reports, and perhaps any income for services
rendered was collected into a Dean's discretionary fund, to be used for
general staff and student activities.

The Wertheimer annual report is significant in two other respects,
the first being the parochial nature of his intentions, and the second, an
absence of any of the research topics being investigated by the Faculty
which might have been of industrial interest. This second issue changed
completely when Andrew Robertson took over as Permanent Dean in
1929; gone was the list of tests carried out for industry, and in came
quite extensive details of the research programmes of academic staff.
But if we go back to the occasion of Robertson's appointment in 1919,
we find in the advertisement details

> It is however, the intention of the Society and the Dean to
> encourage the Professor to keep in touch with the Engineering
> Industry, by undertaking private practice in so far as this will
> not in *their* (my italics) opinion interfere with his work for the
> University.

The italicised word in the above may remind the reader (Chap. 4) that it was Wertheimer's intention as Dean in 1919 that 'their' meant himself (i.e. he thought he <u>was</u> the Society of Merchant Venturers for this purpose), which caused the Vice-Chancellor, Isambard Owen, to object to Wertheimer's personal rule over the Faculty, controlling all the activities of the academic staff, including any outside work.

The quotation does not specifically say that the Professor was encouraged to augment his university salary through such activities, but the words 'private practice' surely indicate that this was the case, and the autobiography[8] of Sir Bernard Crossland (Ref. 21) is most illuminating in this regard. Robertson's strut-testing machine, which he had designed himself and used in the research which earned him his FRS, was also used in 1924 to help Sir Ralph Freeman (of the consulting engineers, Freeman Fox and Partners) in his design of the Sydney Harbour Bridge, and this would have required Robertson to act professionally outside his university contract, and to be paid personally for doing so. Crossland also records in some detail Robertson's early work on improvements to industrial knitting machines, which he had been familiar with from his boyhood in Lancashire, and, with assistance from Prof. John Morrison (Chap. 16), the creation of an entirely new concept for such machines, which in Crossland's presence at a textile trade fair in 1946 '. . . ran incredibly smoothly and produced cloth four times as fast as its nearest competitor. As a consequence FNL Ltd. set up a new factory at Burton-on-Trent to manufacture the machine and its associated equipment.' Crossland does not record what the financial consequences of this invention were to its creator, to his department, or to the University, but we may be sure that Robertson would have been happy to see his successors following in his own image. Indeed, we know that Rawcliffe developed the PAM (Pole-Amplitude Modulation) motors through contracts with Metropolitan-Vickers, and according to one of his younger colleagues became modestly rich in doing so! He was certainly able to own a town-house in Clifton and a small country estate at North Nibley in Gloucestershire! But we also know that he, like the other post-war Professors, used their professional activities to support their research by

---

[8] Bernard Crossland, FRS, joined the Mechanical Engineering Department in 1946 and worked with Robertson in both research and private practice. He became Professor of Mechanical Engineering at Queen's University, Belfast in 1958, and received his knighthood for his chairmanship of the Enquiry into the King's Cross Tube Station fire.

providing financial support for research students and assistants. Pugsley, for example, funded his structural stability research group on contracts with the Aluminium Development Association for the study of the use of aluminium in railway carriages, and British Shipbuilders for ship and submarine design. Additions to salary from the routine testing of materials by more junior colleagues did take place – with permission but certainly without encouragement! And those of us whose fundamental research was of immediate use to industry were informed that any financial reward for 'outside work' which we were invited to carry out, must not exceed one-third of our university salary – this appeared to be a Faculty rule! With reluctance – or so it was said – the University finance officer did allow Heads of Departments to assemble 'discretionary funds', to be used for attendance by themselves or their colleagues at conferences or technical meetings. As far as is known to the author, consultancy was the limit to which an academic in the Faculty was allowed to become involved with industry; further commitment, such as directorships or ownership of companies was not considered to be consistent with essential academic freedoms and responsibilities.

In all that has been recorded above, there remains an uncertainty about the financial arrangements when 'outside work' involved the use of University facilities, as at the start of the Faculty when testing of materials for local firms was a major activity, and when Andrew Robertson used his strut-testing machine for the Sydney Harbour Bridge studies. Until 1985 this issue appears to have rested with the Head of Department, but in that year the Faculty acquired the earthquake shaking table, and with it a responsibility, imposed by the SERC, to satisfy the needs of UK industry by research of course, but also by testing, particularly for the nuclear industry. The solution was to create a company wholly owned by the University (Bristol Earthquake Engineering Laboratory Ltd – BEELAB) for which interaction with industry was formally arranged within quality assurance regulations, and all categories of income and expenditure precisely defined, including payments to all grades of staff who had taken part. In more recent times, it has become common practice to organise all 'outside work' in this way, sometimes through the creation of a University 'Spin-Out' company.

10.5.2 A More Formal Management of 'Outside Work'
It is not known precisely when the University introduced its Code of Conduct for members of staff, but until 1997 these contained rules

concerning 'outside work'. Whether payment was made to the individual or to a University discretionary fund, permission was required if the work involved more than 20 hours *per year* (my italics). It was clearly time for a fresh look at the issues involved, which led to a lengthy paper from the Pro-Vice-Chancellor, Martin Partington, Professor of Law. It attempted to introduce clear definitions of the issues involved, but it only succeeded in raising many serious objections throughout the University. The Vice-Chancellor's method of dealing with these was to appoint a group of three Professors, one each from Law (Ruth Annand), Chemistry (Selby Knox) and Engineering (David Muir-Wood), to gather and assess these disparate views before making their own report. This they did in June 2001, suggesting that a 'minimalist' route should be taken, which encouraged outside consultancy in an active manner, with suitable constraints on the associated risks. What they meant by this was the status quo ante, letting Heads of Departments and Deans continue to ensure that primary duties of research, teaching and administration were not being compromised. In doing so they had taken note of solutions to this issue which had been made in similar UK universities, leading them to make the specific proposal that there should be freedom to undertake personal consultancy up to a total of two days per month without seeking permission; beyond this, permission should be sought from the Dean, and that 'this should not be unreasonably withheld'.

One important factor leading to the acceptance of this approach was the growing contribution being made by RED – the Research and Enterprise Development Service (see below), which had been given the task of actively promoting consultancies and in giving general advice on best practice. In a document 'Policy on Outside Work', the latest of several versions appearing in 2005, RED listed 24 such activities, ranging from external examining to holding executive directorships. And in a companion guide, 'Undertaking Consulting – A Guide for Academics' which appeared in February 2008, in addition to dealing with risk and legal obligations, it gave specific advice on financial aspects relating to both consulting through the University, and the range of charge-out rates for different grades of staff when acting independently.

10.5.3 RED – The Research and Enterprise Development Service

The Dearing Report, published in July 1997 and discussed earlier in this chapter, made several recommendations relating to what is now referred to as the Intellectual Property (IP) of universities. Since its membership contained several vice-chancellors, it is not surprising that – also in July 1997 – the CVCP should distribute to universities a document with the title 'Costing, Pricing and Valuing Research and Other Products'(Ref. 19), and neither is it a likely coincidence that in June 1998 Sir John Kingman should address Senate in the following terms:

> We should therefore continue to be academically ambitious, but if at the same time we are commercially naïve we will not succeed. Financial astuteness and a business-like approach to generating income and managing resources do not threaten research but are the foundations on which any credible and lasting research strategy is built. In order to assist you in this I have appointed Dr. Paul O'prey to a new post of Director of Research Development Services.

There was at this time a Research Support and Industrial Liaison Office (RSILO), and Dr O'prey was to provide a vehicle for collaboration between this, the finance office, the personnel office and the development office. The Deputy Head of the RSILO at this time, Tom Hockaday, was also the University's IP manager, and it was he who was commissioned to produce a business plan for a new University-owned company, Bristol Innovations Ltd (BIL), which had the task of overseeing the exploitation of the University's IP. The plan required the establishment of an IP Management Advisory Panel, to be chaired by a Pro-Vice-Chancellor, with Hockaday as the head of the IP Management Unit, to be funded by BIL through a subsidy payment from the University in its early years.

This somewhat tortuous approach to the important matter of managing IP was quickly clarified by bringing all activities together in 1999 in the Research and Enterprise Development Office (RED). The Government had responded to Dearing (Rec. 40) by its funding of a 'Science Enterprise Challenge', and the University was one of eight winners in the competition, being awarded £2.6m for the creation of the Bristol Enterprise Centre, which became part of RED. An expressed aim of the Centre was to ensure that students in science and engineering (and eventually across the whole university) would be '. . . exposed to the experience and challenge of entrepreneurship, were trained in enterprise skills and fired with the entrepreneurial spirit as an integral part of their studies.'

No doubt there were many in the Faculty who regarded training in 'skills' to be a by-product of their academic duties, but others accepted the challenge with sufficient enthusiasm to create two unit courses (Chap. 8.11), one called 'The Entrepreneurial Option', and the other 'Building Your New Business Venture'. These units attracted students from many parts of the University, but after initial success, enthusiasm waned and the two units were coalesced into one.

Another venture resulting from the Government's positive response to the Dearing Report was an initiative by HEFCE in its 'Higher Education Innovation Fund (HEIF), and it is this which provides the basic financial support for RED, obtained in a two-year competitive cycle. Its other funding comes from a consequence of research activities in the Faculty as elsewhere in the University, being concentrated into fewer, and very much larger areas. Now, the administrative process of obtaining large grants and contracts, and managing them once obtained, is no longer seen as the proper function of the individual researcher – even if he/she had the necessary experience to do so – and it is here that the experience of RED personnel has become necessary. In 2008 it employed ten 'Project Managers', five of whom were responsible for Faculty projects, two of which can be given as examples. The AQUATEST programme aims to deliver a low-cost water test that can be used widely in developing countries by non-experts. Its funding, of £6.8m, comes mostly from the Melinda and Bill Gates Foundation, with the Faculty's Director of the University's Water and Health Research Centre, Prof. Stephen Gundry, leading an international collaborative partnership which includes the Berkeley campus of the University of California, the University of Cape Town and the Royal College of Surgeons of Ireland. In the UK, the University of Southampton and the Health Protection Agency are partners, and in the University itself more than 30 academic staff, in 13 different departments, are involved.

The second Faculty example of the value of RED is the creation of the 'Advanced Composites Centre for Innovation and Science (ACCIS) led by Prof. Michael Wisnom. Not only does this bring together research in composite materials across the whole university, but RED was the catalyst for causing two aerospace companies to integrate their own composite research activities with it; these were Rolls Royce and Smiths Aerospace, the latter also involving the University of Oxford.

Chapter 11

Aeronautical and Aerospace Engineering

A Summary of Significant Events

1945   Bristol Aeroplane Co. (BAC) endows the Sir George White Chair of Aeronautical Engineering in memory of its Founder and first Chairman. It promises £6000 p.a. for 10 years. Arthur Roderick Collar is appointed to the Chair at a salary of £1150 p.a.

1946   Joseph Black appointed Lect. after working at deHaviland and RAE Farnborough; he becomes Sen. Lect. in 1955. In 1960 he moves to the Bristol CAT, which later becomes the Univ. of Bath, at which he is appointed Head of the School of Engineering. He died in 2003.
Camb. Univ. Press reprints *Elementary Matrices – and Some Applications to Dynamics and Differential Equations* by R. A. Frazer, W. J. Duncan and A. R. Collar. It was first published in 1938 and was re-published seven times.

1947   Eileen Watkins appointed to provide computational support for staff and postgraduates.
3' 6" Open-jet wind tunnel installed in MVTC.
Collar awarded G. Taylor Gold Medal for paper on 'The expanding domain of Aeroelasticity'. He is a UK delegate to the Commonwealth Advisory Aero. Res. Council and travels to Australia by flying-boat.

1948   Tom V. Lawson from English Electric appointed Lect.; becomes Sen. Lect. in 1962, and Reader in 1973 in the field of Industrial Aerodynamics. He takes part-time early retirement in 1980 to successfully build a boundary-layer wind tunnel in the Dept. as a commercial enterprise serving the UK Construction Industry. He finally retired in 1996.

First PhD awarded to Rupert Trail-Nash.

1951   Dept. allocated £83 (*sic*) for 'books, periodicals and binding' for each of the next two years.

1953   Miss Geraldyne Smith donates 1000 BAC 5% preference shares to fund entrance and research scholarships for BAC-nominated students.

1954   RAF asks Dept. to give a course on 'Principles of High-speed Flight' for its senior officers. Its success results in a repeat in 1955.

1955   Black is awarded PhD for 'Swept-back Wings-Liquid film Technique for Boundary Layers'.

1956   R. Peter Boswell from Hawker-Siddeley appointed Lect.; promoted Sen. Lect. in 1971 and retires in 1985.

Collar awarded Orville Wright prize by Roy. Aero. Soc.

1958   Jeffrey Tinkler appointed Lect.; becomes Sen. Lect. in 1966 but shortly after takes a post at the Univ. of Manitoba.

1959   Collar and Black organise a Colston Research Society Symposium on Hypersonics; of the 100 participants, 30 are from overseas.

David H. Lloyd appointed Lect., and became Sen. Lect. in 1989. He took partial retirement in 1995 for a 2-year period.

Boswell (with Houghton) publishes a textbook, *Further Aerodynamics for Engineering Students.*

1960   John Flower, from BAC Guided Weapons, appointed Lect., and Sen. Lect. in 1973.

Installation of supersonic wind-tunnel from RARDE Fort Halstead – originally Busemann's tunnel for development of V-rockets in Germany.

Installation and commissioning of Low-Density wind tunnel by Lloyd.

1963   Collar is elected President of Roy. Aero. Soc.

'Slush-drag Rig' developed by research student Rod Barrett for studies related to Munich air disaster (in which most of the Manchester United F.C. were killed). Collar on Enquiry Committee. Rig used later in collaboration with British Aerospace to design spray deflectors for preventing engine ingestion of water on Concorde.

1964   Collar awarded CBE.

Alan Simpson appointed Lect. after research on cable oscillations; promoted Reader in 1972 and was awarded a Personal Chair in 1985. He took early retirement in 1988.

Wind tunnel designed and constructed by Flower and Gerald Avison for icing studies on Concorde.

1965   Collar elected to Roy. Soc. and awarded Gold Medal of Roy. Aero. Soc.

Dick Sawyer appointed Lect.

1966   Brian Hunt appointed Lect. He becomes Reader in 1979, but leaves shortly after for a post with Northrop in California.

1968   Collar is appointed Acting Vice-Chancellor on death of John Harris. He very successfully steers the Univ. through student unrest, and initiates important management changes.

Lawson becomes Acting HoD.

David Birdsall from Canada, and John Webber from RMCS Shrivenham appointed Lects. The former becomes Sen. Lect. in 1990, HoD in 1999 and takes early retirement in 2002. The latter becomes Sen. Lect. in 1984 and takes part-time early retirement in 1995.

1969   The Univ. honours Collar by the award of an Hon. LlD.

Rod Barret appointed Lect., and Sen. Lect. in 1987; he retired in 2003.

1971   The Univ. of Bath awards Collar an Hon. DSc.

1973   Collar retires and is granted Hon. Fellowship of the Roy. Aero. Soc.

Lew Crabtree from RAE (Aerodynamics) is appointed to Sir George White Chair and HoD.

1975   Collar awarded Hon. Fellowship of the AIAA, USA.

1976   Collar awarded Hon. DSc from Cranfield Institute of Technology.

1978   The first Conference on Unmanned Air Vehicle Systems (UAVS) is held. It has been repeated at approximately 2-year intervals ever since.

1979   Low turbulence wind tunnel, designed by Rod Barrett, opened.

Bob Stirling appointed Lect. and joins Alan Simpson on MOD contract in Aircraft Dynamics.

1980   Roger Moses appointed as Faculty Director of Continuing Education. He became Sen. Lect. in 1990 and retired in 2004. His significant contribution was the groundwork for the EuroMEng.

Supersonic tunnel dismantled; German parts given to Industrial Museum.

1985   Collar dies, aged 77. The Univ. had awarded him its Hon. Fellowship, but he died in the week before he was due to receive it.

Simpson invited to be HoD. The book *Matrices and Engineering Dynamics* which he wrote jointly with Collar was published shortly after the latter's death.

David Cowling appointed Lect.

1986   Martin Lowson, from Westlands, appointed Prof. and HoD.

Dept. changes name to Aerospace Eng. but degree name remains the same.

In the first RAE (and in the three which followed in 1989, 1992 and 1996), Aero. and Mech. Eng. are assessed as one Unit; they are awarded Grade 4 (see Appdx. 6 for Grading criteria).

1987   Peter Bunniss appointed Res. Fellow to carry out studies on wind-power, flow visualisation and helicopter noise. He became Sen. Lect. in 2007.

Michael Wisnom, previously Technical Director (Europe) for SDRC Engineering Services, joins as Lect. He is promoted Reader in 1992 and is awarded a Personal Chair in 1995.

Major changes in teaching of aircraft design; Ted Talbot – former Chief Engineer for BAe Airbus, and senior design staff from Filton initiate a 'design modification' exercise (complete aircraft) for third-year students.

1988   Stirling leaves Dept. to found his own Research Company – 'Stirling Dynamics'.

Steve Fiddes – Superintendent of Basic Aerodynamics at RAE Farnborough – joins Dept. as Sen. Lect. He is awarded a Personal Chair in 1997, but leaves to form a spin-out company concerned with aerodynamics of fast cars and yachts.

Barratt leads a major European programme of research on Laminar Flow Technology in which 25 universities and national research establishments involved.

Ian Farrow from Cranfield appointed Lect. and joins Structures/Materials Group. He became Sen. Lect. in 2005.

Gordon Breeze leaves Rolls Royce Derby to join the Dept. as an Industrial Dynamics consultant.

Steve Fiddes installs a £130,000 Titan super-computer for CFD developments.

Nicola Everett – from Oxford – joins as Lect., and becomes Sen. Lect. in 1997. She resigns to take a post at the Univ. of Nottingham in 2001.

Dept. receives a £322,000 SERC award for a 3D fibre-optic Laser Doppler Anemometer; this revolutionises fluid dynamics studies in the Dept. It is a portable facility, with 3-axis traversing, and is useable in three Deptl. locations and externally for consultancy support.

1989   Paul Brinson, from Westland Helicopters, joins as Lect. and initiates Avionics Curricula. He returns to Westlands in 1997.

Dept. receives a legacy of £4000 with which to establish 'The Roderick Collar Prizes', to be awarded to second-year students in the Dept. who have excelled in both their studies and in some activity which has been to the benefit of society.

For the RAE (still jointly with Mech.Eng), a grade 4 is awarded. The three research themes are Aerodynamics and Fluid Mechanics, Structures and Materials, and Control and Systems. The total value of research contracts reaches £1.2m.

1990   Raymond Hale joins from the Royal Navy (Education) to set-up the Aerospace Integrated Graduate Development Scheme (IGDS) in collaboration with UWE. It presents week-long modules for employees in industry and is the only Aero. IGDS in the UK.

Nicola Everett starts involvement with Chemistry Dept for research on growing diamonds on metallic surfaces.

Farrow develops improved manufacturing techniques for thermoplastics, based on a high-frequency hydraulic press.

Webber awarded a Bristol DEng.

Mark Hempsall joins as Lect. from BAe Stevenage (Space). He became Sen. Lect. in 2006 and resigned in 2008.

Nick Lieven, from Imperial College, is appointed Sen. Lect. to lead Dynamics Group. He is promoted to Reader in 1998, and to a Personal Chair in 2002.

The University ILO congratulates the Dept. on its award of 8 BRITE/EURAM European research awards.

1991   Lowson is elected to the Fellowship of Engineering.

John Webber is awarded the Roy. Aero. Soc. prize for the best paper on Structures in 1991.

1992   New course entitled Avionic Systems, made up of IGDS modules in aeronautics, electronics and computer science, and developed primarily by Brinson.

Steve Macqueen joins to become the Avionics Officer.

First undergraduate intake on special World Bank scheme for Indonesian Students.

Cowling leaves to become partner in Stirling Dynamics (see 1988), which now employs 24 recent graduates on advanced dynamics contracts.

Mark Lowenberg arrives from S. Africa and is appointed Lect; he becomes Sen. Lect. in 2003 and HoD in 2007.

For the 1992 RAE the 4A grade is awarded, the 'A' signifying that all academic staff were declared as 'research active'. Space Technology appears as an additional research theme, and attention is drawn to the investment of more than £1m in 3D fibre-optics Laser-Doppler anemometry.

In terms of EU contracts, the Dept. was the UK leader.

1993   First group of undergraduates spend one-year in a European university in the MEng degree.

David Johnson joins Dept. as Lect. to support development of the Avionics degree. He moves to BAE Systems in 1997.

1994   Installation of B-727 simulator, a gift from Dan-Air.

Dave Balmford, Chief Scientist, Westland Helicopters, joins as Visiting Industrial Professor to insert helicopter theory into final-year option.

Fiddes granted a 1-year research sabbatical.

With EPSRC funding, Lowson initiates the Advanced Transport Group. Its remit is to consider the principal features of a new personal transport system for the next century.

1995   Following early retirement from Rolls Royce (Patchway), Sandy Mitchell (a former student) becomes Academic Administrator. He retires in 2006.

First graduates in Avionics.

Authority given for full-time taught MSc using the IGDS modules and an industry-based project. It became very successful, and in 2001 it had 183 students.

Abhijit Guha appointed Lect. and becomes Sen. Lect. in 2003. He and Mitchell accept responsibility for the new Fluid Mechanics and Thermodynamics course which Mech. Eng. previously provided.

Kevin Potter joins as a Sen. Res. Fell.; he is appointed Sen. Lect. in 2002 and Reader in 2006.

Lieven granted a 1-year sabbatical to concentrate on research.

EPSRC provides funds for Lowson's 'Urban Light Transport' (ULTRA) System, which in 2000 becomes a University spin-out company ATS – Advanced Transport Systems Ltd., attracting more than £3m from industry and Govt.

1996   Former students celebrate The Department's Jubilee year.

A grade 4B is awarded in the RAE; the 'B' signifying that 80–94% of staff had been declared as 'research active'.

Aerodynamics research now includes high performance sailing vessels and racing cars, together with wind-turbines – all at full-scale. One half of research funding is now from the EU, and involves more than 30 EU Univs. and Companies.

1997   Ian Bond joins Dept. from Univ. of Reading; he is appointed Sen. Lect. in 2005 and Reader in 2007.

Former student (BEng 1991, PhD 1996), Tom Ellenrieder, appointed Lect. in Control.

1998   Paul Weaver arrives from Cambridge to teach Structures and Materials. He is appointed Sen. Lect. in 2003 and Reader in Lightweight Structure in 2005.

Dr Nick Cook (BRE, Garston) appointed part-time Prof. of Industrial Aerodynamics. He resigns in 2003.

1999   Lowson appointed Emeritus Prof. of Advanced Transport and is invited to demonstrate his ATS concept in the Millennium Dome.

Chris Allen (BEng 1985, PhD 1993) appointed Lect. in Aerodynamics; becomes Reader in 2003, and Prof. of Computational Aerodynamics in 2007.

Dept. joins Faculty team in the BLADE proposal (Chap. 10.3), which is successful in obtaining £15m from the HEFCE and £5m from the Univ.

2000   Lowenberg awarded a 1-year sabbatical for research. He and Mario di Bernardo join with David Stoten (Mech.Eng.) in obtaining an EPSRC award for 'Adaptive control of systems with non-linear and chaotic dynamics – with application to flight dynamics'.

2001   Weaver awarded a 5-year EPSRC Advanced Research Fellowship in 'Structural Efficiency via Material Anisotropy'.

GKN–Westland funds a Res. Fellow to study 'The interaction between contra-rotating rotors'.

For the RAE, a grade 4B was obtained; the research presented was divided into two themes – Aerodynamics and Structures, with subdivision of the first into computational and experimental aspects; in the second the subdivisions were composite materials and structural dynamics.

2002   Lieven becomes HoD.

Mike Friswell from Swansea appointed to the Sir George White Chair.

Hale retires as IGDS coordinator, and is replaced by Wendy Fowles-Sweet from Rolls Royce.

Doug Greenwell arrives from Quinetiq to be Reader in Experimental Aerodynamics.

Steve Hallett appointed Lect. in Aerospace Structures and is promoted to Sen. Lect. in 2007.

Hua-Xin Peng from Oxford appointed Lect. in Materials; he became Sen. Lect. in 2006.

Bond awarded Univ. Research Fellowship and Leverhume Fellowship.

Potter awarded the Peter Allard Silver Medal (Roy. Aero. Soc.) for practical achievements in the use of composite materials.

2003   Allen is awarded an SERC 5-year Advanced Research Fellowship for 'Unsteady Aerodynamic Modelling Tools for Rotary-wing Aerodynamic and Aeroelastic Optimisation'.

Dorian Jones (PhD 1995) and Ann Gaitonde (PhD 1992) jointly appointed to a Lectureship in Aerodynamics – the first such joint-post in the Faculty. In 2007 they were jointly promoted to Sen. Lect.

Sondipon Adhikari arrives from Cambridge to a Lectureship in Dynamics.

Tom Richardson (MEng 1998, PhD 2002) appointed Lect. in Flight Mechanics.

2004   Bond and Adhikari awarded 5-year EPSRC Advanced Research Fellowships for 'Exploiting Functional Fibres in Advanced

Composite Materials', and 'Safety-based Optimal Design in Structural Dynamics', respectively.

Neil Taylor (MEng 1998, PhD 2002) appointed Lect. in Aerodynamics.

Arthur Richards appointed Lect. in Dynamics and Control.

Steve Burrows appointed Lect.

With the opening of the BLADE facilities, a 2-storey test-hall becomes available for studies on aerospace structures with computer simulation using the sub-structuring approach. It is the only dedicated rotorcraft experimental vibration facilities in the UK, and immediately obtains a £1m investment from Augusta Westland for research in this area.

Lieven obtains a £400k DTI award for 'Wireless intelligent sensing'.

Friswell leads a Faculty team to receive a £1.94m Marie Curie award for 'Design of Morphing Aircraft'.

Wisnom obtains a £120k award to study 'High strain-rate effects in notched composites'.

Wisnom and Friswell collaborate with Mech.Eng. on a 3-year project 'Acoustic emission for damage detection', funded jointly by EPSRC, Airbus, Rolls-Royce and BNFL.

2005   Fabrizio Scarpa, Chrystal Remillat and Asikin Isikveren appointed Lect. The first of these became Reader in 2007, and the last, Sen. Lect. in the same year.

Greenwell leaves to a post in City Univ.

University Review Panel congratulates the HoD (Lieven) on progress, and recommends a professorial appointment in Fluid Dynamics.

Total research funding now in excess of £5m – £270k per research-active staff.

The spin-out company ATS (see 1995) obtains £7.5m from British Airports Authority to build a pilot project at Heathrow's new Terminal 5.

Dept. obtains a £120k award for research on 'Efficient aero-elastic design for flexible aircraft'.

2006   Deaths of Crabtree and Boswell.

With Mech.Eng., Dept. co-sponsors the Colston Res. Soc. symposium on 'Adaptive Structures'; published in book form by Wiley in 2007.

Elements of the unique IGDS, offered with UWE (see 1990 and 1995), now form the teaching component of an EPSRC EngD programme 'Aerospace Design, Manufacture and Management'.

2007   Lieven becomes Dean, and Lowenburg HoD.

Panagiotis Margaris and Guiliano Allegri appointed Lects.

Prof. David Ewins – from Imperial College – becomes Director of BLADE.

Launch of ACCIS (Advanced Composite Centre for Innovation and Science) with Wisnom as Director. Based in the Faculty, but with contributions from Science and Medicine, it has strong industrial input from Airbus, Rolls-Royce, GE Aviation and other companies.

2008   For the RAE submission there are now four themes – Multifunctional Composites and Novel Microstructures; Design, Analysis and Failure; Intelligent Structures; and Composites Processing and Characteristics.

Since the last RAE in 2001, 67% of the current academic staff have been appointed, and 50 PhDs awarded; there are currently 57 PhD students and 32 Res. Assts. The outcome of the RAE was joint second place (with Sheffield) in the national list, after Imperial College.

The EPSRC awards ACCIS £6m for a Doctoral Training Centre in which to expand its research and postgraduate education.

Chapter 12

Civil Engineering

A Summary of Significant Events

1909   Robert M. Ferrier, formerly Prof. of Engineering at UBC, becomes the first Prof. of Civil Engineering. He is joined by Frank J. Broadbent (Lect.) and J. Lees (Asst. Lect.). Broadbent resigns in 1915 and is replaced by H. M. Bennett as Temp. Lect. Aside from teaching, Ferrier's main contribution was in testing materials for local industries. Broadbent was the researcher, with the wide interests indicated below.

1911   Alfred J. Sutton-Pippard graduated. In 1914 he obtained an MSc, followed by a DSc in 1920, and in 1928 he became Prof. and HoD.

1912   G. Britton appointed Lect.
Broadbent invents a new apparatus for measuring the impact of water jets.
Reginald E. Stradling graduates. During the Second World War he was Chief Scientific Adviser to the Home Office. He was elected to the Roy. Soc. in 1943 and received a knighthood in 1949.

1913   Ferrier's address as President of the Engineering Society is 'Progress in Engineering'.
Broadbent publishes a paper in the magazine *Engineering* on 'Statical torque tests on a Pelton wheel'.

1915   In the Proc. of the Royal Astronomical Soc., Broadbent publishes 'A method for calculating longitude from observations on the moon'.

1920   W. E. Francis appointed Lect.

1923   Francis joins with Ferrier in studies for the Building Res. Station (DSIR) on the properties of building materials.

1928   Sutton-Pippard replaces Ferrier as HoD. He publishes a book *Strain-energy Methods in Stress Analysis*. Already a member of the Aeronautical Res. Ctte. and Adviser to the Building Res. Station he and Francis begin research on stresses in wire wheels. He was elected to the Roy. Soc. in 1953.

1931   Field survey courses introduced into undergraduate courses; first in Ashton Park, later in various places in the City, including Durdham Downs, the Univ. Sports Ground and Bristol Airport.

1932   William N. Elgood appointed Lect.; Pippard obtains a grant for him from the Steel Structures Res. Ctte. (SSRC) to start research on the bending of plates.

1933   Pippard moves to Imperial College and is replaced by John F. Baker, Chief Technical Officer to the SSRC, of which both Pippard and Andrew Robertson are members. Baker establishes first soil mechanics laboratory in the Dept.

1934   Baker publishes 'An investigation of the stress distribution in a number of 3-storey steel building frames' as Reports of the SSRC.

A. J. Ockleston (one of Baker's Research students) is engaged on an experimental investigation on the effects of earthquakes on steel-framed buildings. (In view of the Dept's later major interest in earthquake engineering (see 1982) it is a pity that no publications ensued.)

1936   Pippard and Baker publish *The Analysis of Engineering Structures*, a book which was reprinted five times in the original edition, and appeared in its fourth edition in 1968.

1937   Baker directs Elgood in research on the stability of loose soil slopes.

Ernest F. Gibbs appointed Lect.; becomes Sen. Lect in 1946 and retires in 1970. He assists the City Engineer in the design of the new drainage system for the City. During 1946–55 he was responsible for the move from Unity St. to Queen's Building. He was also Chairman of the Athletic Grounds Ctte.

1939   Baker seconded to Ministry of Home Security during 1939–45 war. He designed the 'Morrison' air-raid shelter and was awarded the OBE.

Gibbs appointed Regional Technical Officer in the Min. of Home Security – to inspect damage caused by enemy action. During the war academic staff were either in the Forces (Elgood was a Lt.Col. in the Royal Engineers), or spent the vacations advising at Bristol Aero. Co.

1944   Baker moves to Cambridge, and is replaced by Alfred. G Pugsley from RAE, Farnborough, whose war-time work on the torsional stability of aircraft wings had been of the greatest significance.

1945   Peter B. Morice awarded 'best student' prize. After working at the Cement and Concrete Assoc., he was, at the age of 31, appointed to the Chair of Civil Eng. at the Univ of Southampton, and became FREng. in 1989.

1946   Elgood develops facilities for laboratory and field work in soil mechanics, allowing the subject to become part of the degree course for the first time. A. H. Chilver awarded 'best student' prize (see 1950).

1947   Dept. becomes an approved 'testhouse' for the Min. of Supply and the Air Registration Board; 70 firms used the facilities during the first two years.

1948   Pugsley organises Colston Research Society Symposium on 'Engineering Structures'; more than 100 scientists and engineers attended.

1949   M. Stuart G. Cullimore appointed Lect. He was awarded a Merchant Venturers Entrance Scholarship to the study in the Dept. in 1938 and graduated in 1941. He served in the Army for the remainder of the war, and returned to obtain a PhD under Pugsley on 'Light alloy structural sections in torsion'. He became Sen. Lect. in 1961 and Reader in 1972. He retired in 1990 and died in 2007.

1950   Pugsley takes over from Andrew Robertson as a Trustee of the Clifton Suspension Bridge Trust and carries out a detailed study of its structural health. Other members of the Dept. accepted this role until 2006, after which work on the bridge was carried out by the Dept. on an as-needed basis.

   A. Henry Chilver appointed Asst. Lect.; Anthony R. Flint and John B. Caldwell become Res. Fellows. This trio of Pugsley's research students all achieved eminence; Chilver was elected to the Roy. Soc. and became Vice-Chancellor of the Univ of

Cranfield, and later Lord Chilver of that place; Flint became Reader at Imperial College in 1957 and then founded his own firm of Consulting Engineers which established an international reputation for long-span bridges; and Caldwell became Prof. of Naval Architecture at the Univ. of Newcastle. All three became FREng.

1951   Pugsley elected to the Royal Society, and appointed Chairman of the Aero. Res. Council.

1953   Pugsley, as PVC, initiates discussions with the Royal Western Academy on the merger of its School of Architecture with the Univ.

1955   Stuart Armstrong, Malcolm Holmes and William J. Larnach appointed Lects. The first took a post at the Univ. of Sydney in 1958; the second became Prof. of Civil Eng. at the Univ. of Aston in 1959, and the third remained in the Dept., becoming Sen. Lect. in 1965.

Philip S. Bulson appointed Lect., but resigned in the same year. He also was one of Pugsley's research students to achieve eminence as the Director of the Military Vehicles Res. Estab. at Christchurch, Hants., and as a Prof. at the Univ. of Southampton.

1956   Pugsley becomes Knight Batchelor.

Roy T. Severn appointed Lect., and awarded a Telford Premium of the ICE for research at Imperial College connected with the Dokan dam in Iraq. He became the Dept's first Reader in 1965, Prof. and HoD in 1968. He was Dean twice (1971–4 and 1991–4) and PVC in 1981–4. He retired in 2005.

Gibbs builds scale model of Bath Avon for flood alleviation in Bath; research which was to be repeated 50 years later on a different scale model.

1957   Pugsley elected President of IStructE. He publishes *The Theory of Suspension Bridges* a book which has a second edition in 1968.

The Aluminium Dev. Assoc. and British Shipbuilders provide Pugsley with funds for research.

Dept. sponsors an extra-mural course in 'Traffic Engineering'; 60 attendees.

1958   Michael C. Quick appointed Jun. Fell.; promoted to Lect. in 1960, but resigned in 1962 to take a post at the Univ. of British Columbia, Canada.

Severn accepts Membership of the ICE Arch Dams Committee and starts research on earthquake effects with Peter R. Taylor using physical models and finite element methods of analysis. British Transport Commission provides funds, and Min. of Supply provide specially large, thin, tubes for Pugsley's impact research on railway coaches.

1959   Robert Park (from Christchurch, N. Zealand) appointed Lect. to teach and develop research in concrete structures. He returned to a Chair in Christchurch in 1965, and subsequently became a world authority on seismic design of concrete structures.

1960   Alan G. Davenport appointed Res. Fell. His work (with Pugsley) on wind loading on structures earned him a Chair at the Univ. of Western Ontario, Canada, where he built a boundary-layer wind tunnel and an international reputation in wind engineering.

1961   Brian S. Smith appointed Lect.; in 1964 he moved to the Univ. of Southampton, and then to Canada.

The UGC approves Pugsley's scheme for bringing the RWA School of Arch. into the Univ. with special links to the Dept.

1963   Pugsley becomes acting VC due to illness of Sir Philip Morris. Thomas L. Shaw appointed Assist. Lect.; he became Lect. in 1964 and Sen. Lect. in 1980, but resigned to take a post in industry.

1964   Richard G. Redwood (the best student of his year in 1956) appointed Lect.; he resigned in 1965 to take a post at McGill Univ. in Canada.

1965   Keith A. Upton appointed Lect., but resigned in 1968. Michael H. R. Godley appointed Asst. Lect.; promoted to Lect. in 1967 but resigned in 1968.

1966   Norman F. Richards and Roland G. Morgan appointed Lects. Richards was a former student, who worked in the Colonial Service before returning to the Dept. He became Sen. Tutor in Eng. Management in 1971. Morgan took leave to stand for Parliament as Liberal Party candidate at a number of national elections but did not succeed. His research interests were in marine concrete. He retired in 2000.

Pugsley publishes a book *The Safety of Structures.*

1967   Structural Dynamics Group (SDG) awarded SRC grant for construction of a synchronised system of 6 electromagnetic exciters for dynamic testing of laboratory models.

1968   Pugsley retires, but becomes a member of the Ronan Point Enquiry (London high-rise building collapse due to a domestic gas explosion).

Dept. joins with Geology and Botany to obtain NERC grant for a study of the Severn Estuary.

William J. Smith appointed Asst. Lect., Lect. in 1970, and Sen. Lect. in 1984. He retired in 2006 and died shortly after.

1969   David I. Blockley appointed Lect.; he became Reader in 1982 and was awarded the first Personal Chair in the Dept. in 1989. He became HoD in 1990 and Dean in 1994. He retired in 2006.

Roy Dungar appointed Jun. Fell., and became Lect. in 1970. He resigned in 1971 to work for Motorcolumbus in Switzerland.

1971   Robert H. J. Sellin and Norman J. Woodman appointed Lects. Sellin became Sen. Lect. in 1980, Reader in 1984 and was awarded a Personal Chair in Hydraulics in 1992. He was HoD in 1995–7 and retired in 2000.

Woodman became Sen. Lect. in 2001 and retired in 2003.

SRC provides funds for the SDG to purchase a large electrodynamic exciter for creating a 'slip-table' for earthquake engineering research.

1972   Dept. joins with Depts. of Geology and Botany to organise the 1978 Colston Res. Soc. Symp. on 'Tidal Power and Estuary Management', and edit a book of the Proceedings.

1973   Severn and Blockley begin a 12-year surveillance, at 6-monthly intervals, of floor-levels in Gloucester Cathedral to monitor subsidence due to drainage caused by flood management in the River Severn.

1976   Building Research Establishment (BRE) awards £34k to SDG for construction of 4 sychronised eccentric-mass exciters for dynamic testing of prototype structures. The design and construction was carried out by J. Parry (Mech. Eng.) and G. Read (Elect. Eng). The first application was on the National Westminster Bank building in London, then under construction.

1978   SDG uses eccentric-mass exciters to study dams in the UK, after which it accepts an invitation, jointly with the BRE, to carry out dynamic tests on the Emosson and Contra dams in Switzerland.

1979 Alfred Pugsley is awarded the James Alfred Ewing Medal by the ICE and Roy. Soc. jointly, for 'Special meritorious contribution to the science of engineering in the field of research' (This medal had been awarded in 1943 to R. E. Stradling, in 1952 to J. F. Baker, and in 1963 to A. J. S. Pippard; the first being a former student and the last two being former HoD).

David F. T. Nash and Peter Waldron appointed Lects. The former became Sen. Lect. in 1995; the latter was appointed Reader in 1988 before moving to the Univ. of Sheffield as its Prof. of Civil Eng. in 1992.

1980 The UGC appoints Severn to its Technology Sub-Committee (1980–88); he takes part in establishing the structure of the first Research Assessment Exercise (RAE) for UK universities.

1981 Severn becomes a Pro-Vice-Chancellor. With two BRE colleagues he receives the Telford Gold medal of the ICE for tests on dams, and is elected to the Fellowship of Engineering.

1982 John P. Davis and Colin I. Robertson appointed Lects. Davis becomes Sen. Lect. in 1993 and Reader in Hydroinformatics in 1996, but Robertson soon returns to a post in industry.

SERC awards £473k for creation of a 6-axis shaking table for earthquake engineering research; they also make an award for the creation of a soil mechanics Field Centre on a soft-clay site at Bothkennar in Scotland for which David Nash was the Principal Proposer. It also funds the creation of a large open-channel facility at HRS Wallingford – Robert Sellin being one of the proposers.

Severn becomes Chairman of the 'Seismic Effects' Committee of the International Commission on Large Dams.

Bulson, Caldwell and Severn organise a celebration of Pugsley's 80th birthday and edit a presentation book, *Engineering Structures – Developments in the Twentieth Century*.

1983 Blockley appointed to the 'Limestone Mines Executive' (Chairman Sir Edward Parkes) to advise on risk and safety of abandoned workings.

1984 The SDG collaborates with Dagless (Elect. Eng.) and Thomas (Comp.Sci.) to develop a computer-vision system for measuring non-contact displacement of structures in the laboratory or in the field.

1985   The Earthquake Eng. Res. Centre (EERC) is created, and studies the dynamic behaviour of the Humber suspension bridge. It is invited to perform similar studies on the two Bosporus suspension bridges in Turkey, and also by Stretto di Messina Spa. to perform further Humber tests, together with the Politechnico di Milano, to assist in design studies for the proposed Straits of Messina crossing.

1986   In the first RAE the Dept is awarded the highest grade of 5.
Severn becomes a Member of the Engineering Board of the SERC.

1987   Colin A. Taylor, a recent research student, is appointed Lect.; he becomes Reader in 1992 and is given a Personal Chair in E'quake Eng. in 2004. He became HoD in 2008.
Anthony R. Blakeborough appointed Lect.; he moves to the Univ. of Oxford in 1992 and is awarded a Personal Chair there in 2008.
The 4-year MEng with study in Continental Europe is offered for the first time.

1988   Blockley awarded Telford Gold medal by ICE, and the George Stephenson medal by IStructE.
John Loveless appointed Lect. on the closure of the Civil Eng. Dept. at King's College, London. He is promoted Sen. Lect. in 1995 and takes part-time early retirement in 2008.
Waldron and Davis awarded £100k by the SERC Marine Tech. Directorate to study the on-site behaviour of a novel concrete armour unit for breakwaters, which Waldron had designed.
Martin L. Lings appointed Lect. promoted to Sen. Lect in 1997.

1989   EERC obtains an SERC Teaching Company Scheme with Ove Arup for experimental validation of seismic design software for structures and foundations.
BEELAB (Bristol Earthquake and Engineering Laboratory) created as a University-owned Company to exploit the commercial use of the shaking table. A unique QA system is developed based on nuclear engineering standards (ISO 9000 and BS 5750).
In the RAE the Dept. again received Grade 5; its five research themes were structural dynamics, polymer-induced drag reduction and open channel flow, civil eng. systems - risk and safety, fatigue in bridge members, and full-scale measurement of concrete structures.

1990   Severn elected President of ICE, and is invited to join the UK National Ctte. for the UN's IDNDR – International Decade for Natural Disaster Reduction; he chairs its Structures Ctte.

Blockley introduces a Civil Engineering Systems course into the 4-year degree.

EERC receives a grant of £100k from SERC for its vibration studies on as-built structures.

1991   EERC is awarded EU contracts with laboratories in Greece, Italy, Portugal and France to study shaking tables behaviour. The MCS algorithm of Prof. Stoten (Mech. Eng.) provides enhanced performance resulting in the EERC becoming Coordinator during 1993–2004 of the EU's programmes in experimental aspects of earthquake engineering, which bring to the Dept. researchers from all parts of Europe.

Sellin becomes Editor of the ICE Journal 'Water, Maritime and Energy'.

1992   Deptl. Industrial Advisory Committee established.

The Dept. acts as 'guinea-pig' for proposed University 'Academic Performance Assessments' of Departments.

For the RAE the Dept continues to receive Grade 5; to the 1989 research topics it now adds its involvement in Geotechnics. Research income now averages £37k p.a. per academic staff.

Sellin receives £130k SERC award for research 'On overbank flows in meandering channels with sediment transport' at the new facility at HRS Wallingford (see 1982).

Blockley is elected to the Roy. Acad. of Engineering. He is also awarded £110k in partnership with Nuclear Electric, HSE and Halcrow for risk assessment studies.

Severn receives CBE for services to the Construction Industry, and an Hon. DEng from the Black Sea Univ. in Turkey.

1993   Dr Sally Heslop joins as Lect.; she becomes Sen. Teaching Fellow in 2004.

Dept. joins with BRE in a £67k contract to monitor floor-loading in the new Faculty building on Woodland Rd (later, the MVB).

Davis is awarded £220k by the UK Technology Foresight Programme jointly with the Petroleum Institute for a study of 'Uncertainty in estimating oil reservoir reserves'.

A dinner is held to celebrate Pugsley's 90th birthday; friends and colleagues present him with written 'Recollection and Tributes', edited by John Caldwell. (see 1950).

1994   Adam J. Crewe appointed Lect.; he becomes Sen. Lect in 2005 and is responsible for the E'quake Lab. in BLADE.

Taylor is appointed UK representative on the ICOLD Seismic Effects Ctte. He is awarded £192k by Nuclear Electric to develop methods for assessing the strength of unreinforced masonry walls under seismic loading.

Severn is awarded £27k by Nuclear Electric for a desk study of methods used to design nuclear reactor cores against seismic loads.

1995   Severn retires and is replaced by Prof. David Muir-Wood, Prof. of Civil Eng. at Glasgow Univ; he becomes HoD in 1997.

Muir-Wood is awarded a 2-year Roy. Soc. Industrial Fellowship for secondment to the Babtie Group of Consulting Engrs.

Safety Systems Research Centre (SSRC) established with Dr. John May as Director.

1996   Wendel Sebastian from Jesus Coll. Camb. is appointed Lect., and Sen. Lect. in 2006.

For the TQA, Dept receives 22 out of 24; its two deficiencies are in 'teaching resources' and 'quality resources and enhancement'.

The EERC obtains permission to monitor the cable behaviour of the Second Severn Crossing bridge; John Macdonald is the R.A. responsible.

Davis introduces an 'e-learning' package for teaching hydraulics to students in Civil Eng. and Eng. Maths.

1997   Water and Environmental Management Research Centre (WEMRC) founded with industrial funding. Prof. Ian D. Cluckie, from Univ. of Salford is appointed to the Chair, and is also elected to the Roy. Acad. of Engineering.

Dawei Han appointed Lect.; Sen. Lect. in 2005 and Reader in 2008.

Jim Hall appointed Lect.; becomes Reader in Civil Eng. Systems in 2004 but resigns to take a Chair at Newcastle Univ. in Earth Systems Eng.

Taylor receives JREI award for equipment for the E'quake. Lab.

1998   Sir Alfred Pugsley dies.

Blockley appointed Non-Executive Director of Bristol Water plc.

Severn awarded the James Alfred Ewing medal by ICE and Roy. Soc. (see 1979).

Hall awarded 5-year RAE Research Fellowship in 'Handling uncertainty in Civil Engineering Systems'.

Crewe and Wendy Daniell receive £29k from SERC in its 'Public Understanding of Science' programme to develop IDEERS (Introducing and Demonstrating E'quake Eng. Research in Schools). After success in the UK, the British Council invites it to visit Japan and Taiwan; in 2004, with sponsorship from Industry, it held international competitions in these countries.

John M. H. Macdonald appointed Lect., and becomes Sen. Lect. in 2006.

Muir-Wood is elected to the Roy. Acad. of Engineering. He conceives and organises a Roy. Soc. meeting on 'Mechanics of granular materials in engineering and earth sciences'; he edits the resulting volume.

1999 Muir-Wood receives £54k (50% JREI funding) to construct a 'cubical cell for studying anisotropy of soil stiffness.'

EERC and the ACT laboratory in Mech. Eng. develop real-time dynamic sub-structuring, allowing large systems to be tested experimentally in the laboratory.

Sellin awarded an ICE Telford premium for paper on 'River channel hydraulics.'

An optional course in 'Engineering Architecture' is introduced for second-year students and proves to be very popular.

2000 Alan Feest joins Dept. as Lect. on closure of Dept. of Continuing Education.

Erdin Ibraim appointed Lect; promoted Sen. Lect in 2006.

The book 'Doing it Differently' by Blockley and Godfrey (see 2006) receives the CIOB gold medal.

Muir-Wood, Nash and Lings receive two EPSRC awards – 'Effects of damage to soil-structure on the small-strain stiffness of Bothkennar clay', and 'Shear strain response of sand at small/medium strain levels under multiaxial stress states.'

2001 Blockley elected President of IStructE.

Ian Duncan appointed Sen. Teaching Fellow and takes charge of the 'Engineering Architecture' course.

For the RAE, Dept receives 5*C – indicating that not all staff were submitted for assessment. Conforming with Univ. policy, research has been organised into 3 groups – systems, structures and dynamics; geomechanics; water and environmental management.

Highways Agency funds Macdonald to continue monitoring the Second Severn Crossing; he make significant discoveries on rain/wind effects.

ICE awards Hall both the George Stephenson medal and the Palmer Prize.

2002 Jitendra Agarwal appointed Lect., and Sen. Lect. in 2006.

Severn receives the Manuel Rocha award from LNEC Lisbon for 'relevant achievements in research in civil engineering'.

EERC receives £10k from EPSRC to provide an exhibit at the Roy. Soc. Summer Exhibition – 'Shaken but not Stirred'.

2003 Muir-Wood becomes Dean.

Nicholas A. Alexander appointed Lect. and Sen. Lect. in 2006.

Safety Systems Research Centre (SSRC, Director, John May) transfers from Computer Science. It is no longer supported exclusively by industry, but has EPSRC funding.

2004 Dept. experiences University's 'Teaching Quality Assessment' process and is commended for the excellent organisation of undergraduate programmes and for the support of MEng students whilst studying abroad.

H.M. the Queen opens BLADE and visits the E'quake Lab.

MacDonald receives £315 EPSRC award for 'Non-linear dynamic cable behaviour and cable-deck interaction of cable-stayed bridges.'

Lings and Dietz (RA) are awarded the ICE Crampton Prize for 'An improved direct shear apparatus for sand.'

Sally Heslop becomes Dean of Graduate Studies.

2005 Taylor and Crewe obtain £313k EPSRC award (with Oxford and Cambridge) for creation of a UK Network for Earthquake Engineering Simulation (UKNEES) to parallel the US NEES organisation and to collaborate with it.

BLADE obtains an 'Access Grid Node' for tele-conferencing.

Sellin awarded ICE Telford Gold medal.

Muir-Wood presents 20th Bjerrum lecture 'The magic of sands'.

Two former students, C. Huxley-Reynard and George Gibberd set up their own company – Marine Current Turbines Ltd.

Smith and Stephen Bright (p.g. student) awarded the ISE Henry Adams medal for a paper 'Fatigue performance of laser-welded steel bridge decks'.

2006   Patrick Godfrey (from Halcrow) appointed part-time Professor of Systems Engineering and coordinator of the EngD degree.

MacDonald awarded 5-year EPSRC Research Fellowship in 'Aeroelastic and Non-linear Dynamic Interaction of Slender Structures'.

With support from EPSRC and W. S. Atkins, Dept celebrates 200th anniversary of I. K. Brunel's birth by sponsoring a 'Clifton Bridge' competition; 1000 entries, the winner being an Egyptian engineer.

2007   The Engineering Management Group (Peter Strachan, Stephen Gundry, Mohammed Wanous, and Kenneth Iwugo) merges with the Dept. In 2001 it had obtained an EU contract for study of the water-supply problems of sub-Saharan Africa, and this led to Gundry obtaining a £6.5m international water research programme funded by the Bill and Melanie Gates Foundation.

Suby Battacharya and Miguel Rico-Ramirez appointed Lects.

Dr. Andy Hughes (Atkins Ltd. and Vice-Chairman of the International Ctte. on Large Dams) appointed Visiting Ind. Prof.

SSRC is a member of the new Faculty Systems Performance Centre which links with British Energy to maintain safely, to optimise operational performance, and to support future nuclear generation initiatives. It is also a key part of two EPSRC EngD programmes, one in Systems, and the other in Nuclear Eng.

2008   Gundry awarded a Personal Chair in Enterprise and Entrepreneurship.

SSRC obtains EC funding to design and assure the next generation of European avionics systems; 37 partners are involved.

For the RAE there are now 3 groups; Dynamics Eng., Environmental Management and Health, Systems and Safety. It achieves 8th position in the UK listing.

Cluckie resigns to become PVC at the Univ. of Swansea.

Macdonald publishes an explanation of the Millennium Bridge vibrations in Proc. Roy. Soc. (A).

EPSRC awards £6m for development of an 'Industrial Doctorate Centre in Systems Engineering' to stimulate research and postgraduate education.

Muir-Wood resigns and returns to Scotland.

Chapter 13

Computer Science

A Summary of Significant Events

1988   Senate Minutes of 30 Jan 1989 recommend that 'Formal steps be taken to transfer responsibility for the Computer Science (CS) and Computer Science with Mathematics courses from the Faculty of Science to the Faculty of Engineering; transfer to take place for the 1989 session.

Prof. David Warren, creator of the Warren Abstract Machine joins from Manchester Univ. He resigned in 1995.

1989   Prof. Michael H. Rogers joins the Faculty as HoD of Computer Science. With him are Fraser G. Duncan (Reader), Geoffrey N. Lance (Sen. Res. Fell.), and Lecturers C. J. Burgess, M. Flower, C. F. Gibbins, I. J. Holyer, E. Lewis, D. F. Paddon, B. R. Stonebridge and B. T. Thomas.

Dr. D. J. Lloyd appointed Prof. of Computer Science.

For the 1989 RAE submission the leading research topic was collaboration by Thomas with Electronic Eng. in producing a computer-vision system for cancer treatment. Grade 3 was awarded. (For the first RAE in 1986 Grade 2 had been obtained; for the grading system see Appdx. 6).

Alan Chalmers and John Gallagher appointed Lects.; the latter becomes Sen. Lect in 1997 and resigns in 2002.

1990   A set of Macintosh computers introduced for first-year teaching.

Steve Gregory appointed Lect.

1991   Hewlett-Packard donates a number of HP 6800 workstations to create a colour graphics laboratory.

1992   Through collaboration with Elect. Eng and Eng. Maths, and an infusion of new academic staff, for the RAE the Dept. was able to point to success in logic programming, parallel computing, machine intelligence, and computer vision. An Advanced Computing Research Centre had also been created. A Grade 4A was awarded.

1994   An MSc course in Advanced Computing (character animation) is offered for the first time, to be followed in the next session by another on Advanced Computing (global computing and multimedia). Further MSc courses were added, so that by 2008 there were six, the two already mentioned, plus 'Advanced Computing (internet technologies with multimedia)', 'Advanced Computing (internet technologies with security)', 'Advanced Computing (machine learning and data mining)', 'Advanced Microelectronics-System Eng.', and 'Computer Science-for graduates from other disciplines.'

1995   David May FRS (from ST Microelectronics and the inventor of the Transputer, a small microprocessor) appointed Prof. and HoD.

Barry Thomas awarded a Personal Chair in Vision Systems. He died in 2003.

The Safety Systems Res. Centre receives £500k sponsorship from Nuclear Electric, the Civil Aviation Authority, Railtrack and Lloyd's Register.

A Deptl. Handbook is produced for students, setting down course and examinations structure, and progression details. The ability of students to produce their own projects has been found to lead to a number of new companies.

Colin Dalton and Hendrik Muller appointed Lect.; the latter becomes Reader in 2000 and resigns in 2007.

1996   For the RAE, the Dept was able to declare a research achievement in depth, based on the 3 themes of Computer Architecture, Declarative Systems, and Multimedia. It was also able to list the many contacts which had been established with local Industries, such as Hewlett-Packard and the Filton Aerospace Companies. It was awarded a Grade 5A.

Andrew Calway and Neill Campbell appointed. The former becomes Sen. Lect. in 2002, and the latter Reader in 2007.

1997   Majid Mirmehdi appointed Prof. in Image Analysis.

Pieter Flach appointed Sen. Lect.; he becomes Prof. in 2003.

1998   Kirstin Eder appointed Lect.; becomes Sen. Lect. in 2006.

1999   Nigel Smart (from Hewlett-Packard) appointed Reader; he is awarded a Personal Chair in 2004 and becomes HoD in 2007.

Richard Jozsa appointed Prof.

2000   Collaboration with Hewlett-Packard produces a 'Cyberjacket' – with sensors built into the lining.

2001   For the RAE seven research areas were recognised – Quantum Computing, Cryptography, Machine Learning, System Design and Verification, Digital Media, Languages and Architecture, and Mobile Computing. A Grade 5A was awarded.

Five Companies sponsor the establishment of a 'Research Centre in IT and Law', with main areas for study being e-commerce, cybercrime, digital rights management, and privacy.

Dhiraj Pradhan appointed Prof.

Timothy Kovacs appointed Lect.; becomes Sen. Lect. in 2007.

A Deptl. Information Officer is appointed to deal with development and maintenance of an interactive system allowing staff and students to communicate via web interfaces.

2003   DTI and a number of Companies provide £12.6m for a University Innovation Centre embracing Communication, Computing and Content (the 3C Res. Centre).

Dan Page, Waterio Mayol-Cuevas, James Marshall and Mike Fraser appointed Lects; the last of them becomes Sen. Lect. in 2006.

Rafal Bogacz appointed Lect.; becomes Sen Lect. in 2007.

2004   *The Times* ranks Dept. 3rd in UK behind Cambridge and York.

Kirsten Cater and Raphael Clifford appointed Lects.

2005   Dept. moves from Queen's Building to the Merchant Venturers Building (MVB).

Colin Burgess retires; he made a significant contribution to the planning of the Dept's accommodation needs in the MVB.

Christopher Melhuish appointed Prof.

Aram Harrow and Erik Reinhard appointed Lects; the latter becomes Sen. Lect. in 2007.

2006   David Cliff appointed to a Chair. Julian Gough is appointed Reader.

Bogdan Warinschi appointed Lect.

2007   The spin-out company XMOS (based initially on the work of an u.g. student Alan Dixon) secures £16m venture-capital funding. It is developing a new generation of microprocessor chips, described as 'Software Designed Silicon'.

David Cliff is the Director of an EPSRC-funded (£9m) research and education initiative in the science and engineering of 'Large-scale Complex IT Systems'. The universities of Leeds, Oxford, York and St Andrews are also involved. A £5m contribution from industry is expected.

The Exabyte Information Research Theme is awarded three RCUK Fellowships.

A Computer Science Society is founded in the Dept.

Julian Gough appointed Reader, and Oliver Ray Res. Fell.

Srinam Subramanian and Simon Hollis appointed Lects.

2008   Quantum Computing is the Dept's intended contribution to the new Univ. Centre for Nanotechnology.

Nello Christiani begins a large-scale research programme on automatic pattern analysis of the complex global media system, combining artificial intelligence techniques from machine translation to machine learning.

For the RAE the submission consisted of five main themes – Architecture and Design, Digital Media, Foundations, Intelligent Systems, Personal Systems – each one involving a set of more specific topics. The result was 18th place in the UK ranking, from 81 Units of Assessment.

Chapter 14

Electrical and Electronic Engineering

A Summary of Significant Events

1909   David Robertson, formerly of the MVTC, becomes the first Professor of Electrical Engineering in the Faculty. The Lecturers who joined him were J. Williams, H. Stanley, A. Smith, G. Mogg and T. W. O'Connor-Parnell.

1911   Robertson continues to be very active in research, publishing 11 papers (of unspecified length) in either *The Electrician* or the IEE Journal, on such subjects as 'Electrical Potential in Alternating Magnetic Fields', 'Power Transformers', 'Electric Meters on Variable Loads', and 'Migration of Ions in Electrolysis'. He also invented an electric chronometer.

1912   F. N. Tipton, A. E. White, W. J. Cockshott, C. E. Morgan, R. J. Hill and G. E. Stockall all appointed Lects.

1916   Robertson publishes an IEE paper on 'Scientific Research and our Future Supply of Energy'.

1918   Paul Dirac (subsequently Prof. of Mathematics at Cambridge and a Nobel Prize winner in Physics) enters the Dept. as an undergraduate.

1919   Dept. occupies new laboratories in Orchard St.

1920   P. A. Mainstone, S. Holmes and I. Williams appointed Lects; W. S. Palmer, A. V. Baker and J. H. Riley appointed Asst. Lects.

1923   S. Pugh appointed Asst. Lect.
       Holmes wins an IEE Premium for a paper 'A vector treatment of long transmission lines.'

1924   Robertson carries out research on a mechanism for striking the hours on the Great Bell of the Univ. involving investigations on driving magnets and the time-keeping of pendulums. With Messrs. Gent and Co. of Leicester, he develops a new electric clock having 'silent action'.

1925   E. A. Pink and B. J. Smith appointed Asst. Lects.

Miss C. O'Parnell presents Dept. with a galvanometer and slide-wire bridge made by her late brother the Rev. T. W. O'Connor-Parnell (see 1909 above), a former student and teacher at the MVTC.

1926   F. B. Wrighton, H. R. Beasant, and F. R. Gresswell appointed Asst. Lects.

With financial support from Messrs Gent and the Colston Research Society, Robertson and A. W. Hirst study the effects of temperature, pressure, humidity, and the shape of the bob, on the erratic behaviour of the pendulum of the Univ. clock. They conclude that 'barometric errors' are the cause, and publish a series of articles in the 'Horological Journal'.

1928   Only David Robertson and S. Holmes recorded as being members of the Academic Staff.

1931   Frederick de la C. Chard appointed Asst. Lect; promoted to Sen. Lect. in 1955, and in 1971 takes a Chair at the Univ. of Khartoum.

1932   Robertson continues to publish papers on pendulums, but to illustrate his wide engineering interests, publishes 'Static Balance of a Shaft with Skew Stiffness' in 'The Engineer', and 'Whirling of a Journal in a Sleeve Bearing' in the Philosophical Magazine.

Chard publishes 'Selenium Cells – an investigation of certain properties of importance in connection with television', and 'Films in Three Dimensions', both in the Discovery Magazine.

1935   Chard advises the RSPCA Humane Slaughtering Ctte. in connection with electro-lethally (*sic*) of animals.

1939   Chard enlists in the Army and becomes Chief Signals Officer, Southern Command, eventually with the rank of Major.

1941   David Robertson dies.

1943   Gordon Hindle Rawcliffe, aged 34, and from the Univ. of Aberdeen, replaces Robertson as HoD.

1946   Raymond L. Russell, from British Thomson-Houston, appointed Lect.; promoted to Reader in 1955, and later moves to a Chair at the Univ. of Newcastle.

1947   Rawcliffe appointed to the National Council on Education and Industry. With C. M. Brownsey he publishes 'A rectification meter for the indication and measurement of phase angle' in the IEE Journal. Holmes retires, and receives an MSc degree in recognition of his 28 years' service.

1949   Rawcliffe becomes Dean. He was Dean for the second time in 1963.

       R. L. Cannell appointed Asst. Lect.; promoted to Lect. in 1951.

1952   A. Ray Billings and Joseph E. Brown appointed Lects.

1954   Dept. moves from MVTC to Queen's Building. Metro-Vick awards Dept. £500 p.a. for 7 years for research on dynamo-electric equipment. English Electric Co. and General Electric Co. provide bursaries for research students, and donate 'machinery of substantial value'.

1955   William Fong joins Dept. as Jun. Res. Fellow.; appointed Res. Fellow in 1965, and Reader in 1969. In conjunction with The Regional Council for Further Education, the Dept. holds a 4-day course on 'Cross-field Generators'; 60 people attended.

1956   Harry Sutcliffe appointed Lect.; promoted to Sen. Lect. in 1960.

1957   Ray F. Burbidge appointed Sen. Tutor in Electrical Power, to Lect. in 1958 and Sen. Lect. in 1972.

       L. Garlick appointed Lect.

1958   First paper on PAM (Pole Amplitude Modulation) – for speed changing of induction motors – published by Rawcliffe, Burbidge and Fong. The significant financial success of this invention allowed Rawcliffe to employ a number of Res. Assts. to develop it under his direction.

1959   PAM motors exhibited for the first time by Metro-Vick at the Earl's Court Exhibition

1960   W. G. Johnston appointed Lect.

       Alec R. W. Broadway appointed Jun. Fellow; becomes Res. Fellow in 1970 and Reader in 1984. He retires in 1997 and becomes Visiting Fellow.

1961   David J. Storey appointed Lect.; promoted to Sen. Lect. in 1964.

1962   Peter L. Moreton and Edward G. Salthouse appointed Lect.

1963   Tom H. O'Dell appointed Lect.

       Brian Michael Bird (an u.g. student during 1952–5) appointed Res. Asst. with funding from CEGB; becomes Reader in 1969, and Prof. and HoD in 1979. He was Dean during 1985–8 and PVC during 1990–3.

1965   W. E. M. McCoy appointed Lect.

1966   D. Rothwell and Geoffrey A. L. Reed appointed Lect; Reed promoted to Sen. Lect in 1984.

1967   Laurie Burbridge appointed Res. Fellow; becomes Lect. in 1980 and later moves to the Univ. of Exeter.

1968   At the request of the Faculty Board, Senate establishes a Committee to consider an appointment to the Chair of Electronic Engineering.
Alan Jones appointed Lect.

1970   Sidney R. Bowes appointed Lect.; awarded a Personal Chair in 1985 and retires in 2007.

1971   Kenneth. F. Sander (Sen. Tutor, Trin. Coll., Camb.) appointed Prof. of Electronic Eng.

1972   Rawcliffe elected to Roy. Soc. and Vice-President of the IEE.
Anthony J. Copping, John Gowar, and Paul A. Llyn appointed Lects. The first of these became Sen. Lect. in 1990, and Deputy HoD during 1997–2002; he retired in 2002. Gowar retired in 2000.

1973   The Dept. changes its name to 'Electrical and Electronic Engineering.'

1974   Rawcliffe retires. Sander becomes Dean.

1976   Rawcliffe becomes a Founding Fellow of the Fellowship of Engineering.

1978   Sander and Reed publish *Transmission and Propagation of Electromagnetic Waves*, Cambridge Univ. Press.

1979   The Imperial Group offer funding for a Chair in Microelectronics for an initial period of 5 years; the UGC complements this with recurrent funding of £25k p.a. and an initial equipment grant of £21k. The SRC and DTI indicate their willingness to support the Chair through grants and contracts.

1980   Rawcliffe dies.

1982   Eric L. Dagless appointed to the Imperial Group Chair of Microelectronics Engineering. He was HoD during 1985–1988, and then Dean until 1991.
Martin J. P. Bolton and Michael H. Barton appointed Lects. The latter became Sen. Lect. in 1992 and Deputy HoD in 2003.

1983   In response to the Alvey Directorate's coordination of research in Information Technology, a 3-year degree in Computer Systems Eng. is established in collaboration with Comp. Sci. and Eng. Maths.; this is the first inter-departmental degree programme in the Faculty.

Dagless co-founder of ECAD, now known as Europractice.

1984   Sander retires.

Dagless invited by Addison Wesley to be editor of its book series *Electronic Systems Engineering.*

Bird elected to the Roy. Acad. of Engineering. He becomes a member of IFIP working group 10.3.

Graham Norton appointed Lect.

1985   Joseph P. McGeehan appointed to a Chair of Communications Engineering and Director of the Research Centre for Communications Research funded by Nokia and Ericsson. The Centre has groups in modularisation and signal processing, antennae and propagation, VSLI for speech processing, and EMC. It has become the leader in narrow-band linear technology.

Andrew Bateman appointed Lect., Reader in 1990 and Prof. of Signal Processing in 1994. He resigned in 1997.

1986   Dagless appointed by DES to the Board of the Microelectronic Support Unit to promote the use of microprocessors in schools. He also became Chairman of IFIP working group 10.3 for 3 years.

In the first RAE the Dept. was awarded Grade 3 (see Appdx.6)

1987   David J. Edwards appointed Lect.

A part-time MSc in Information Engineering is offered jointly with Eng. Maths. and Comp. Sci. Hewlett Packard agree to provide setting-up costs, and with British Aerospace to provide a regular supply of students.

A Univ. Research Centre in Communications is established – the second RC in the Univ.

1988   Duncan A. Grant, J. L. Glover and Christopher. J. Railton appointed Lects. Railton was awarded a Personal Chair in Computational Electromagnetics in 1999.

The Dept. sets out its research plans in three themes – Communications Eng., Computer Systems, and Power Electronics, Actuators and Drives. In the first it aims to maintain its world lead in mobile communications technology; in the second it will be part of an ESPRIT consortium in Home Systems, and in the third it will join with other Depts. in the Faculty in work on 'pulse-width modulation applied to power electronics'. The achievements and plans in these three areas form the basis of the Dept.'s 1989 RAE submission.

1989   Dept. adds a 3-year degree in 'Electronics and Communication Eng.' to the existing three in Electrical Eng., Electronic Eng., and Communication Systems.

For the RAE a Grade 2 was obtained.

Mark Beach and Geoffrey Hinton apptd. Lects. The former becomes Sen. Lect. in 1996, Reader in 1998, and Prof. of Radio Systems Eng. in 2003; the latter becomes Sen. Lect. in 2004.

1990   Univ. 'Campaign for Resource' assists Dept. in negotiation with Toshiba Research Enterprises (Europe) Ltd. to create a Chair in 'Communications Networks'.

Kennington, Edwards and McGeehan awarded IEE Mountbatten Premium for outstanding paper on 'Satellite Communications'.

Dagless coordinates 3 EU TEMPUS projects having 8–15 partners and valued at €1.75m over 7 years.

First intake in the MEng with Study in Continental Europe – the EuroMEng.

Swales, Beach, Edwards and McGeehan awarded IEEE Premium for work on Adaptive Antennas.

Philip Mellor appointed Professor of Electrical Engineering.

1991   McGeehan becomes HoD until 1998.

Dagless awarded Roy. Soc. Leverhulme Trust Sen. Research Fellowship and also demonstrates golf buggy under autonomous control in a line-following experiment.

Swales, Beach, McGeehan and Edwards win IEE prize for best paper on Vehicular Technology.

1992   A 'Prince of Wales Award for Innovation' is received for research on 'speech scrambling', recommended for use by Police Forces and Special Services in the UK. It is licensed to Marconi Secure Radio and sold as MASK.

Dagless, Milford and Thomas (Comp. Sci.) demonstrate first number-plate recognition system for detecting stolen cars.

Schlumberger Industries (France) Premium awarded for pioneering work in radiocommunications.

Motorola Research Foundation (USA) award for contribution to mobile network research.

Fundamental work underpinning wireless LAN Technologies commenced through EU ESPRIT LAURA project.

Dagless initiates a series of TLTP projects (EDEC) over 9 years with £1.5m EPSRC funding.

In the RAE submission major research groups are communications, signal processing, digital systems and image processing, applied electromagnetics, high-speed actuators and robotics, and industrial electronics and control. Over the assessment period its research funding and number of publications per academic per year had increased to £36k and 5.0, respectively. In addition, £300k was earned through small contracts with industry.

These developments resulted in Grade 4A being awarded.

David Bull apptd. Lect., becomes Sen. Lect. in 1993, Reader in 1994, and Prof. of Signal Processing in 1997.

1993   Dept. institutes the K. F. Sander Memorial Prize; and in recognition of his outstanding interest and skill in teaching, the prize is to be awarded to 'the undergraduate having the highest examination marks in final-year honours'.

Bull and Horrocks awarded IEE Ambrose Fleming premium for paper on 'Primitive Operator Digital Filters'.

Dept. are winning finalists (SW-region – Univs./Colleges Division) of the CBI/Toshiba Year for Innovation, for work in broadband linearised amplifiers.

1994   McGeehan elected to Fellowship of The Roy. Acad. of Engineering.

Univ. Senate is 'agreed to support in principle a commercial company allied to the Centre for Communications Research under Profs. McGeehan and Bateman.'

Dagless appointed TQA Assessor in Computer Sci. by Scottish HEFC.

Derrick Holiday apptd. Lect., and Sen. Lect. in 2005.

1995   In the new Merchant Venturers Building 700m^2 is dedicated to a research Laboratory and clean room.

First intake of students in the MSc in Communications Systems and Signal Processing, and in the 4-year MEng.

BT provides £161k 'to develop new radio propagation models and network planning algorithms to further enhance the quality of service of wireless networks' (Birmingham and Oxford are partners).

Dritan Kaleshi apptd. Lect., and Sen. Lect. in 2007.

1996   Conforming to Univ. policy of fewer, but larger research groups, the Dept now has the two 'themes' of Communication Systems and Industrial Electronics. In the first are wireless and optical communication systems, signal and image processing. In the second are motors and motor control, power semiconductors and conditioning, and applied electromagnetics and EMC.

Dept. now has 39 PhD students and 33 Res. Assts. £8million has been donated by companies for research equipment.

Prof. Ian White from Bath Univ. is appointed to a Chair in Optical Communications as a result of 'Campaign for Resource'. Richard Penty accompanies him as Lect.

Dept is a founder member of the DTI Virtual Centre of Excellence in Digital Broadcasting and Multimedia Technology, with Bull as a Director.

Academic and administrative staff move from QB to MVB.

Dagless appointed HEFCE Assessor for TQA in Electrical and Electronic Engineering.

In the RAE a Grade 5A is obtained.

Paul Warr apptd. Lect, and Sen. Lect. in 2007.

1997   McGeehan awarded £1.4m JREI grant for an 'Integrated Communications and Teaching Laboratory'.

Bull et al. awarded IEE premium for paper on 'Scanning conformal near-red microscopy: a new technique for 3D histopathology'.

First demonstration by Nix and Bull of the use of wireless LAN (HyperLan) in providing robust in-home distribution of broadcast digital radio. (An EU FP4 WINHOME project collaboratively with Dassault Electronique).

EPSRC awards Norton £117k for research on 'The trellis structure of algebraic–geometric codes'.

For the TQA Dept. scores 24/24.

Judy Rorinson apptd. Lect.; Sen. Lect. in 2000 and Reader in 2004.

1998   McGeehan becomes Dean until 2003; he is also appointed Managing Director of Toshiba Research Ltd.

With a £12million investment, Toshiba sets up a Telecommunication Research Laboratory (TRL). £1millon provided for a Chair and a Lect.

A University 'spin-out' Company is formed – 'Wireless Systems International Systems Ltd.' – having 120 employees.

Dept. is the first European laboratory to receive direct funding from NTT Japan, for radio measurements in the 5GHz frequency band.

DTI awards £128k to develop education and training in Radio Frequency Engineering; it is supported by 20 companies.

David Redmill and Kevin Morris apptd. Lects.; the latter becomes Sen. Lect. in 2000.

1999   Martin Cryan apptd. Lect.; Sen. Lect in 2001.

2000   White becomes HoD; he coordinates a £3.2million research award in the EPSRC consortium – PHOTON – involving six Univs. and seven Companies.

2001   White resigns to take a Chair at Cambridge, and Bull becomes HoD.

Andrew Nix is awarded a Personal Chair in Wireless Communications Systems.

For the 2001 RAE the Dept. was able to claim international excellence in Wireless Communications and Industrial Electronics. Its publications in quality refereed journals was 11.2 per researcher, compared to 5.7 in 1996; its research grant portfolio amounted to £11.5m – £500 per researcher – compared to £6m in 1996.

Bull and Nix founded a spin-out company – 'Provision Communication Ltd'.

Goodrich Power Systems establish a Research Centre in Electrical Drives.

A Grade 5A is obtained in the RAE.

2002   Dept. becomes a founder member of the £50m Defence Technology Centre in data information fusion. Bull leads the research theme in multidimensional fusion.

Comp. Systems Eng. transfers to Comp. Sci. Dept.

A £4m SRIF award allows relocation and enhancement of the Wireless and Network laboratories to create a new video processing studio in MVB, and refurbishment of the Photonics clean-room in QB.

A legacy from the Arnold family in memory of Charles Leonard Arnold, founder of MK Electric Co., provides postgraduate scholarships.

2003   After identifying synergy between Communications and Creative Media Industries in the S.W. Region, and in collaboration with the Bristol Enterprise Centre, a £12.6m joint venture University Innovation Centre is established, which embraces a convergence between Communications, Computing and Content (3C research). The DTI contributes 66% of the funding and Industry 33%, the companies involved being STMicroelectronics, Thales, Toshiba, QinetiQ and a number of SMEs.

Alistair Munro appointed to the Toshiba Chair in Communications Networks (an externally funded chair).

John Rarity appointed Professor of Optical Communications.

BLADE laboratories provide new space for the Electrical Energy Management Group, supporting research into hybrid and all-electric technologies.

Mike Barton takes over as Deputy HoD.

2004   McGeehan awarded CBE for services to the Communications Industry.

Dagless appointed Faculty Education Director and Undergraduate Dean.

Rarity leads Euro. consortium to win Descartes Prize.

Craddock extends work on microwave radar to medical applications; successful trials on early detection of breast cancer results in establishment of a Univ. spin-out Company – MICRIMA.

Canagarajah awarded personal Chair in Multimedia Signal Processing.

Legacy from Blake family in memory of Leslie Reginald Blake (graduate in 1945) provides undergraduate scholarships and supports special projects.

Angela Doufexi, Alan Achim, Bernard Stark and Jose Nunez-Yanez appointed Lects.

2005   Craddock awarded the J. A. Lodge prize from the IET for outstanding work in Medical Engineering.

2006   Beach appointed HoD.

Dept. joins with Comp. Sci to offer MSc in Advanced Microelectronic Systems Eng.

Dept's research is now structured under the Centre for Communications Research and the Electrical Energy Management Group, with a funding level at £4m p.a. compared with £1.7m p.a. in 2002, making it a major contributor to the University's financial surplus.

Univ. Review Panel comments on Dept's all-round excellence, international reputation and 'vast array of links with industry'. Formation of the Bristol Vision Institute – a unique collaboration of vision scientists in Europe; Bull becomes one of its Directors.

Rarity awarded a Wolfson Merit Award.

Craddock's team receive the IET award for 'Innovation in Electronic Breast Cancer Detection'; clinical trials start in 2007.

Canagarajah becomes Faculty Research Director.

Toshiba provides undergraduate scholarships.

David Drury appointed Lect. within the Energy Management Group.

Tashim Kocak apptd. Sen. Lect.

2007 The 3C Research Group produces important results, including a state-of-the-art wireless testbed (OSIRIS), advanced video coding techniques (ROAM4G), and innovative content-based retrieval algorithms (ICBR).

McGeehan listed in *Who's Who*.

Beach awarded travel grant by Japan Society for promotion of Science to visit Hokkaido Univ.

Bull and Nix develop portable wireless video technology to enhance the viewing experience of spectators at sports events.

Dept. celebrates 25 years of IET accreditation for its u.g. programmes.

2008 Jeremy O'Brien appointed Professorial Res. Fellow.

For the 2008 RAE the Dept. was able to report the following developments since the RAE in 2001:

1. an increase in academic staff from 22 to 32;
2. £6.4m of targeted investment in research infrastructure;
3. a 30% increase in research income, with one-half being from UK and overseas Industry;
4. postgraduate registration up 20% and PhD awards up 40% per year;
5. a 13% increase in journal papers published per year;

6. 64 distinct patent activities, including 24 joint filings with industry and 12 assignments/licences;

7. 2 spin-out companies, 2 affiliated start-ups, and strategic relationships with Toshiba, QinetiQ and Goodrich.

The result of the RAE was 17th place in the listing of 35 Units of Assessment.

An MSc in Communication Systems and Signal Processing is offered.

Chapter 15

Engineering Mathematics

A Summary of Significant Events

1909   Edmund. S. Boulton, formerly of the MVTC, becomes Prof. of Applied Mathematics in the Faculty of Engineering. He occupied this post until 1935 and died in 1946. His accompanying Lecturers are A. Pickering and H. Panter.

1913   G. E. Stockall appointed Lect.

1919   J. W. Minshull, G. A. Stephens, A. J. Seymour, L. Sampson and T. Hoad appointed Asst. Lects.

1921   A. Pickering, A. Smith, M. E. Hobbs, A. Ackroyd, J. F. Murphy and J. H. Haynes appointed Asst. Lects.

1922   A. D. Banfield, W. P. Palmer, G. Pugh and G. M. Tinknell appointed Asst. Lects.

1923   F. B. Wrightson appointed Asst. Lect.

1925   G. M. Hinton, E. W. Gregory and E. K. Perdue appointed Asst. Lects.

        (Author's note. It would appear that the MVTC practice was to appoint a large number of assistant lecturers on short-term, and possibly part-time, contracts. This situation changed after Andrew Robertson had taken over from Wertheimer as Permanent Dean.)

1927   S. Toby Newing appointed Asst. Lect. He became Lect. in 1936, Sen. Lect. in 1955, and did not retire until the mid 1970s.

1931   William Morgan Shepherd appointed Asst. Lect. He was promoted to Lect. in 1935, to Reader in Elasticity in 1943 (only the second Reader in the Faculty), and to the newly created Chair in Theoretical Mechanics in 1959.

1935   Shepherd awarded DSc degree for a series of papers in the theory of elasticity, and replaces Boulton as HoD.

1939   During the Second World War Shepherd and Newing worked at Bristol Aero. Co. during vacations. Shepherd was also an ambulance driver during air-raids on Bristol.

1946   Frank Alan Gaydon appointed Lect. after war service in the Royal Navy; he becomes Sen. Lect. in 1958 and Reader in 1968. From 1952 until his retirement in 1980 he was Instructor Lieut. Commander of the Univ. Naval Contingent at HMS *Flying Fox.*

1949   The Dept. changes its name to 'Theoretical Mechanics' with Shepherd as its HoD.

1952   The Engineering Professors report to Senate 'That they wish to consider in due course the establishment of a Chair in Theoretical Mechanics and for the appointment to the Chair of Dr. W. M. Shepherd'.

1955   Harry Nuttall appointed Lect.; becomes Sen. Lect. in 1964.

1956   Leonard Sowerby appointed Lect.

1965   Jim F. Baldwin appointed Lect.; becomes Reader in 1981 and is given a Personal Chair in Information Technology in 1990. He retired in 2003.
       T. Faulkner appointed Asst. Lect.

1966   Frank P. Sayer appointed Lect.

1967   A. Tony Richardson appointed Asst. Lect.; becomes Lect. in 1970, Sen. Lect. in 1984 and Reader in 1990. He retired in 1998.

1968   Jon H. Sims-Williams appointed Asst. Lect; becomes Lect. in 1972 and Sen. Lect. in 1992. He took part-time retirement in 2005.

1971   Shepherd retires and is replaced by Ronald D. Milne (QMC, London) as Prof. and HoD.
       David T. Bickley appointed Lect., Sen. Lect. in 1987 and retired in 2002.

1972   A. Watson appointed Lect. He soon became seriously ill and died shortly after.

1973   Dept. changes its name from Theoretical Mechanics to Engineering Mathematics. Richard R. Clements appointed Lect.; becomes Sen. Lect. in 1988, Reader in 1993 and Professorial Fellow in 2005 for his exceptional contribution to teaching in the Faculty.

1977 The Engineering Mathematics BSc-degree programme launched with 12 students.

1978 Gordon J. Reece appointed Lect., Sen. Lect. in 1988, and took part-time early retirement in 1997 until 2004.

Jerry H. Wright appointed Lect, Sen. Lect. in 1991. He resigned in 1999.

B. W. Pilsworth appointed Lect, Sen. Lect. in 1993 and resigned in 2002.

1979 Bickley carries out mathematical modelling for the Severn Tidal Barrage project within the University's SABRINA programme funded by NERC.

1980 First graduates in the BSc (Eng. Maths) degree.

Milne publishes his book on *Applied Functional Analysis*.

1983 The Information Technology Research Centre created with Jim Baldwin as Director. Its computer language FRIL becomes commercially available, providing powerful tools for modelling probabilistic and fuzzy worlds.

1984 Ian J. Thompson appointed Lect., but resigns in 1988.

1986 For this first RAE (Research Assessment Exercise) the Dept was a joint 'Cost Centre' with the Univ. Dept. of Mathematics. A Grade 3 was obtained, signifying it 'to be about average in the UK'. (see Appdx. 6)

1988 Trevor P. Martin (R. A. in 1983) appointed Lect., Sen. Lect. in 1996, Reader in 1999 and awarded a Personal Chair in 2005.

1989 Bickley becomes HoD.

For the second RAE the Dept. is a separate 'Unit of Assessment' and is awarded Grade 3 on the basis of research activities which include electrodynamics in collaboration with CRNS in France, dynamic verification of finite element models with the Civil Eng. Dept, and the study of fields in optical wave-guides with the Comms. Eng. Research Centre.

Andrew P. Dixon appointed Computer Officer; transferred to the Faculty in 1999.

Reece apptd. Academic Staff Training Coordinator for the University.

1990 Baldwin awarded the DSc degree and a 5-year EPSRC Senior Research Fellowship in Information Technology. Michael Greenfield appointed as temporary replacement.

1992   Milne retires and is replaced by John Hogan (from Oxford) whose research field is non-linear dynamics and chaos.

In the RAE the Dept. is assessed with Applied Maths. to receive Grade 4A, which indicates that some areas of research are at international level.

I. Colin Campbell appointed Lect.; promoted Reader in 2004.

1993   Alan Champneys appointed Lect., Reader in 1998 and Prof. in 2001. He joins John Hogan in founding the Applied Nonlinear Mathematics (ANM) Group.

1994   Dept. joins the Faculty decision to move to a 4-year MEng. Degree.

Reece produces software for a 'Faculty Student Information System', allowing academic staff to communicate with students by e-mail.

1996   Jerry Wright resigns and is replaced by Mike Barry, first as a Temp. Lect. but promoted to Sen. Lect. in 2003.

Baldwin's Artificial Intelligence Group (Baldwin, Martin, Pilsworth, Campbell and Sims-Williams) develops 'fuzzy logic' and a theory of uncertainty management of knowledge-based systems applied to machine intelligence, decision making and expert systems. This, and the work of the ANM group, results in a Grade 5B in the third RAE – again as part of Applied Maths.

Barry publishes two books, *Foundation Mathematics* and *Mathematics in Engineering and Science* – both co-authored by L. R. Mustoe of L'boro Univ. and published by John Wiley and Sons.

Hogan becomes Editor of the IMA J. of Applied Maths.

1997   Champneys awarded 5-year EPSRC Advanced Fellowship in 'Analysis of localised phenomena and their applications'.

Mario di Bernardo appointed Lect., becomes Reader in 2003 and is appointed to a Personal Chair in 2007.

Jonathan Lawry (PhD student 1994-7) appointed Lect., promoted to Reader in 2004.

1998   Hogan becomes HoD.

Bernd Krauskopf appointed Lect., Reader in 2001 and appointed to a Personal Chair in 2003.

Two u.g. degrees now offered – MEng in Engineering Maths., and MEng in Computational and Experimental Maths.

Dept. receives 23/24 in the HEFCE Teaching Quality Assessment – 'teaching resources' being the deficient item.

1999   Eddie Wilson appointed Lect.; promoted to Reader in 2004 Clements awarded MBE (Military Division) for services to Naval Education.

A third u.g. degree offered – MEng in Mathematics for Intelligent Systems.

2000   The ANM group organise the 52nd Colston Res. Soc. Symposium on 'Non-linear Mathematics and Chaos – where should we go from here?' and edit the proceedings in book form. Bernd Krauskopf designs chaos posters for London Underground stations.

2001   Hinke Osinga appointed Lect.; promoted to Reader in 2005.

Martin Homer appointed Lect.; Sen. Lect. in 2007.

First cohort of 26 students (7 female) enrol for the new 5-year Industry-led, interdisciplinary, Faculty-wide, MEng in Eng.Design. 12 sponsoring Companies taking part. Barry and Sims-Williams are the organisers.

Bernd Krauskopf begins 5-year EPSRC Advanced Fellowship on 'Non-linear Dynamics and Global Bifurcations in Semi-conductor Laser Systems'.

Jonathan Rossiter appointed Lect.; Sen. Lect. in 2007.

Martin awarded BT Senior Research Fellowship.

In the fourth RAE, again with Applied Maths., a Grade 5*A is awarded, indicating that the majority of the research is now at international level.

2002   ANM Group - now the Bristol Centre for Applied Non-linear Mathematics (BCANM). It receives £1m, the largest ever EPSRC award in the field of Applied Mathematics. Its aim is to recruit five, three-year, Res. Assts. over the 5 years of the award to work on specific themes – relation to delays, piece-wise smooth dynamics, numerical methods, spacially extended systems and dynamic substructuring.

2003   Rossiter receives 4-year Roy. Soc. Japanese Fellowship.

2004   Champneys becomes HoD.

Dept. concentrates its u.g. teaching into two MEng courses; the 5-year Eng.Design, and the 4-year Eng.Maths. From 2005, the 4-year degree allows the third year to be taken in an approved overseas university.

The Univ. Review of the Dept. remarks on its uniqueness in the UK, and on its award of 5* in the 2001 RAE. On teaching, it records that ' The Department's reputation has developed from simply offering the teaching of Maths to other Engineering Departments in the Faculty to providing a number of its own, extremely good, undergraduate programmes'.

Osinga and Krauskopf appear 'live' on Channel 4 News, and other worldwide media outlets, for rendering the 'Lorenz Manifold' in crochet.

2005   Dept. celebrates 25th anniversary of the first graduates from its own Eng.Maths. Degree.

Nello Cristianini appointed as Professor of Artificial Intelligence.

Osinga begins 5-year EPSRC Advanced Fellowship on 'Global invariant manifolds, applications, critical boundaries and global bifurcations'.

2006   First graduates from the 5-year Engineering Design degree.

John Terry appointed Lect.; Sen. Lect. in 2007.

Cristianini awarded Wolfson Roy. Soc. Merit Award for his work on informatics and news mining.

BCANM group awarded £1.6m from EPSRC for 'Applied Non-linear Mathematics – Making it Real'. Problems to be studied include epilepsy, rattling gears, biomechanics of walking and hearing, and hybrid testing of complex engineering components.

2007   The Bristol Centre for Complexity Science, led by John Hogan, opened with a £4 million EPSRC award. It will recruit 15 postgraduates annually for one-year training in advanced-level mathematics, statistics and computer science, followed by 3-year programmes of research.

Mario di Bernardo awarded Italian Cavaliere for his unique collaboration between Bristol and Naples.

Campbell receives a 3-year award from Cancer Research UK for 'Use of bio-informatics techniques to find targets of therapeutic interest'.

Dr. Yulia Kyrychko (ex R.A.) now on EPSRC 3-year Research Fellowship 'Challenges of modeling hybrid testing for specially extended systems'.

Tijl de Bie appointed Lect. in Artificial Intelligence.

Caroline Colijn and Krasimira Tsaneva-Atanasova appointed Lects., both working in mathematical physiology, and bringing the number of female permanent academic staff up to three.

Wilson begins 5-year EPSRC Advanced Fellowship in ' A multiscale framework for forecasting highway traffic flows.'

Undergraduate annual intake now in excess of 30.

2008   Luca Giuggiolo and Konstantin Blyuss appointed Lects. in Complexity Science.

David Barton (ex u.g. and p.g.) awarded a Great Western Research Fellowship on 'Power Harvesting'.

Wilson appears twice in TV documentary programmes to explain the phenomenon of 'phantom' traffic jams.

Dept. assessed jointly with Applied Maths in the RAE.; it is ranked 3rd in the UK list, from 46 Units of Assessment.

Chapter 16

Mechanical Engineering (and Mining until 1927)

A Summary of Significant Events

1909   John Munro, formerly of the MVTC, becomes the first Professor of Mechanical and Mining Engineering in the Faculty, and is joined by Lecturers F. R. B. Watson, T. J. Moss-Flowers, G. Britton and F. C. Webber. Watson becomes the Faculty's first Sen. Lect. in 1941.

1911   Watson publishes the Dept's first research paper 'Variation of temperature in the fuel-bed of a suction gas producer', having received support from the Colston Res. Society. His principal interest was internal combustion engines, in which he published a series of papers in the magazine *Engineering*, such as 'Gas engine pistons and exhaust valve temperatures'. He retired in 1945, and died in 1952.

1912   Univ. awards Munro a 'service' degree of MSc, but he did not retire until 1918 after 48 years' service, 39 of them at the MVTC. He died in 1930.

        A. Fisher appointed Asst. Lect.

1914   During the 1914–18 World War there was little research activity. Staff were either in the armed forces or were engaged in activities connected with the war, one of which was classes for workers in munitions factories; by 1915, 304 men and 233 women had been trained. Another duty was instructing Officers of the Royal Flying Corps. in engine maintenance.

        Munro, as President of the 'Engineering Society' presented an Address on 'The building and water supply of ancient Rome'.

1919   Munro is replaced as Prof. and HoD by Andrew Robertson. As a lecturer at Manchester Univ. he was drafted into the Royal Naval Volunteer Reserve (RNVR) during the war and reached the rank of Major; but he actually worked at the Royal Aircraft Establishment (Farnborough) on aircraft structures, particularly wing-struts.
J. W. Minshull and G. A. Stephens appointed Lects.; A. T. Seymour, L. Sampson and T. Hoad appointed Asst. Lects.

1920   R. K. Fry and H. King appointed Asst. Lects.

1921   Robertson constructs a strut-testing machine for use in his research; it is also used in assisting Sir Ralph Freeman with the design of the Sydney Harbour Bridge. (A modified version of this machine was still in use until the BLADE reorganisation of Queen's Building laboratories in 2003.

1922   F. J. Lynes, J. J. C. Benson and W. Williams appointed Asst. Lects.

1923   W. P. Palmer, T. J. A. Rogers, T. Rogers and J. D. Haddon appointed Asst. Lects.

1924   Robertson made Permanent Dean of the Faculty and Principal of the MVTC – even though he was the youngest Professor. ICE awards him the 'Telford Gold Medal' for a paper on 'The Strength of Struts', which contributed to the 'Perry-Robertson' formula and which, with amendments, is still used.

1925   Robertson publishes his experiments on local buckling in thin tubes, causing existing theories to be revised.
A. W. M. Wintle appointed Asst. Lect.

1927   With A. J. Newport, Robertson publishes his research on yielding of mild steel cylinders under internal pressure. This research was the beginning of the Dept's major research theme carried forward by Morrison, Crossland and Parry until 1970.

1928   The academic staff consisted only of Robertson, Watson and Stephens, but the last of these resigned during the session and was replaced by John L. M. Morrison as Assst. Lect. He was promoted Lect. in 1932, Reader in 1945, Prof. in 1945, and HoD in 1946.
A 50-ton testing machine, funded by a UGC grant, but designed by Robertson for compression and bending tests, was installed in the Dept. He published 'The strength of tubular struts' in the Proc. Roy. Soc. and two other papers on the same theme. A year later he extended this research to elastic stability of struts in

torsion, and the influence of tube-length and wall thickness – a topic which was to be developed by Pugsley (Chap.12) 20 years later.

1932   William H. Dearden appointed Lect. in Metallurgy, and becomes Sen. Lect. in 1961.

Robertson becomes a member of the Steel Structures Res. Ctte. of the DSIR and of the Aeronautical Res. Ctte.

1935   Robertson is appointed to the Advisory Council of the DSIR. He also accepts membership of the Clifton Suspension Bridge Trust (CSBT), and conducts an investigation of its structural health. (The role of structural advisor to the CSBT continued to be taken by a member of the Faculty until 2006).

1938   Morrison, having been 'tutored' by Robertson on precise experimental work, produced a series of papers of the fundamental properties of various types of steel (e.g., 'The influence of strain-rates in tension tests') for which he was awarded a DSc degree by the Univ. of Glasgow.

1939   As in the First World War, academic staff worked in local industry during vacations, Morrison and Dearden with the Bristol Aero. Co. Dearden was also a 'welding Inspector' for the Ministry of Aircraft Production. All Faculty laboratories became approved test-houses for the Aeronautical Inspection Dept. By 1943, 70 firms had made use of them.

1940   Robertson elected to the Royal Society.

1941   Robertson becomes a member of Council of the Royal Society. Morrison is a member of the Gun Design Ctte. of the Ministry of Supply.

1943   Robertson becomes a Member of the UGC, and also assists the Ministry of Labour and National Service with the State Bursary scheme for students.

1945   Robertson is elected President of the IMechE.

1946   Morrison replaces Robertson as HoD.

Gordon F. C. Rogers and C. D. Graham appointed Lects. Rogers is promoted Sen. Lect. in 1958 and becomes Prof. of Engineering Thermodynamics in 1964. He retires in 1982.

Bernard Crossland appointed Asst. Lect.; promoted Lect. in 1950 and Sen. Lect. in 1956. In 1958 he accepted a Chair at Queen's Univ. Belfast, and whilst there became Chairman of the Ctte. of Enquiry into the King's Cross tube station fire, for which he received a knighthood. He was elected to the Roy. Soc. in 1979.

1947 Morrison joins the Advisory Council on Scientific Research and Technical Development of the Ministry of Supply; he becomes its Chairman in 1957.

1948 Yon. R. Mayerowitz appointed Asst. Lect., promoted Lect. in 1951 and Sen. Lect. in 1966.

DSIR and ICI fund research on 'The behaviour of thick cylinders subjected to repeated applications of high pressure'.

1949 Robertson persuaded the City of Bristol to buy the MVTC building for use as its own Technical College, allowing him to approach the UGC for funds for a new home for the Faculty. He retires from the Univ. and is made an Honorary Fellow – the highest honour it can bestow.

1950 Morrison joins the Agricultural Res. Council, and Rogers the Gas Turbine Ctte. of the Aeronautical Res. Council.

Archibald E. Russell, a 1924 graduate and now Chief Designer and Director of BAC, is awarded the Gold Medal of the Royal Aero. Soc.

1951 As an indication of the way research was funded at this time, the Dept. received £200 from ICI for electronic equipment, which was typical of the many such small gifts. In 1958, the Dept. received grant of £2000 'for assistance with research projects', and Morrison himself received £335 'for purchase of research equipment'.

1952 A. N. Dickson and W. G. Wood appointed Lects. Both resign in 1955, the former to take a post at AERE, Harwell; the latter to the State Univ. of Pennsylvania.

1955 Morrison, Parry and Crossland awarded the IMechE George Stephenson medal for 'High pressure fatigue research on thick-walled cylinders'.

Thomas D. Eastop appointed Jun. Fellow, and Peter M. Threlfall appointed Lect.

1956 A fatigue testing machine developed in the Dept. is sold to Brown Univ. in the USA.

Jack. A. Bones and John F. C. Parry appointed Lects. Bones becomes Sen. Lect. in 1971, and Parry in 1968.

1957 Morrison awarded CBE for 'Services to Education and Industry'.

Rogers and Mayhew publish *Engineering Thermodynamics* – a book which became a standard text both in the UK and abroad; it was revised and reprinted many times, the last by Mayhew alone after the death of Rogers in 2004.

1958   Michael A. Hollinsworth appointed Lect.; promoted Sen. Lect. in 1976 and retired in 1994.

1959   The Univ. bestows the Hon. Degree of Doctor of Laws on Robertson. Morrison, Crossland and Parry awarded the George Stephenson prize by the IMechE for a paper 'Fatigue under tri-axial stress; development of a testing machine and preliminary results'.

1960   Derek J. Haines appointed Lect.

1962   Colin Andrew appointed Lect.; promoted Reader in 1969; Prof. and HoD in 1971. He resigns to take up in post in industry in 1982, but accepts a Chair at Cambridge in 1985.

1964   Morrison publishes *An Introduction to the Mechanics of Machines*. It was revised and reprinted in 1970 and 1985.
Herbert Saravanamuttoo appointed Lect.; he resigns in 1970 to take a post in Canada.

1965   Start of MSc course in 'Fluid and Thermal Studies for Industry'; this was the Faculty's first 1-year MSc course.

1966   E. Graham Ellison appointed Lect; becomes Reader in 1974 and Prof. and HoD in 1984. He retired in 1990.
William Barker and Peter W. Fitt appointed Lects. Barker died in 1971; Fitt retired in 1990.

1968   Robert D. Adams appointed Lect. He becomes Reader in 1980, Prof. in 1986 and retired in 2005.

1971   Morrison retires and is elected President of IMechE.
Robert Poulter appointed Lect. and Sen. Lect. in 1987.

1972   Brian J. Stone and William J. Plumbridge appointed Lects. The former resigned in 1981 to take a post in Australia; the latter became Reader in 1986, but resigned shortly after.

1977   Andrew Robertson dies.

1983   Rogers publishes a book, *The Nature of Engineering*.

1984   David P. Stoten appointed Lect.; promoted to Reader in Automatic Control in 1989, Prof. in 1994 and HoD during 1998-2004. He obtains the first two SERC Teaching Company Schemes (TCS) in the Faculty, both with Avon Rubber plc, for investigating the properties of extruded rubber, and the implementation of modern concepts in automatic control to mixing, extrusion and other manufacturing processes. He also creates a Faculty Automatic Control laboratory for research on the adaptive control of robotic devices.

Christopher A. McMahon appointed Lect. He becomes Sen. Lect. in 1993, Reader in 1998, and retires in 2002.

1985   John E. Morgan and Stuart Townley appointed Lects. The former becomes Sen. Lect in 1988; the latter becomes Sen. Teaching Fell. in 2007.

Enhanced 3-year BEng and 4-year MEng courses introduced with common first two years; more than half of student cohort sponsored by industry on 1-3-1, or 1-4-1, basis.

Korosh Kodebandehloo (from Imperial College) is awarded a 'new blood' lectureship. He creates the Advanced Manufacturing and Robotics Res. Centre (AMARC) and obtains large EU contracts. But his skill in obtaining these funds was not matched by his ability in dealing with them, resulting in him resigning from the Univ. in 1995.

1986   On the closure of the Dept. of Architecture, three of its staff (Brian Day, Terry Gorman and John Bracey) transfer to the Dept.

For the first RAE the Dept. is assessed jointly with the Aerospace Dept. and a Grade 4 is obtained (Appdx. 6).

1987   R. D. Brooks appointed Sen. Lect. in Eng. Management Studies. Peter N. Brett appointed Lect., Sen. Lect. in 1994, Reader in 2000. He resigns in 2003 to a Chair at the Univ. of Aston.

Institute of Grinding Technology founded.

1988   Martin Pavier appointed Lect., Sen. Lect. in 1999, Reader in 2002 and is awarded a Personal Chair in 2008.

MEng in 'Mech. Eng with Manufacturing Systems' established with earmarked UGC funds; 15 students in first cohort.

D. Smith appointed Lect.; becomes Reader in 1992 and Prof. in 1996.

1989   Stoten publishes his 'Minimal Control Synthesis (MCS)' algorithm. Its use in the Faculty was a principal reason for obtaining funding for the BLADE re-development of Queen's Building.

For the RAE, Dept. reports creation of an Industrial Advisory Board, and that two Companies are providing the salaries of two academic staff. Its research is divided into 3 groups – Materials, Advanced Manufacturing and Automation, and Thermal Studies. Its research contracts amount to £2.2m. Grade 4 is obtained.

1990   Andrew Harrison appointed Lect.; becomes Sen. Lect. in 2000. Kazem Alemzadeh appointed Eng. Officer, Lect. in 2005 and Sen. Lect. in 2007.

The Food Refrigeration and Process Eng. Res. Centre (FRPERC) is affiliated to the Dept.

1992   Giovanni (Joe) L. Quarini, from AEA, Harwell, appointed to ICI/ Zeneca Chair in Process Eng. following success in the 'Campaign for Resource'.

For the RAE, research is now centred on 6 groups – Automatic Control, Composites and Adhesives, Design, Engineering Materials, Robotics, and Thermal Studies. Its research funding per academic staff is £70k p.a. and it receives more EU funding than any other UK Dept. Grade 4A is obtained.

1993   Stoten awarded a Bristol DEng and a 1-year Univ. Res. Fellowship.

Michael Tierney appointed Lect. and becomes Sen. Lect. in 2004. In the TQA assessment, the Dept is adjudged to be 'Excellent' – the highest Category.

1994   The Control Group uses MCS to produce real-time adaptive control of the EERC shaking table. This achievement leads to EC-funded research contracts during the next 10 years, valued at €9 million, in collaboration with the majority of European laboratories in this field, and subsequently with laboratories in the US and Japan.

1995   Quarini and colleagues win £850k MAFF contract for food processing research; 4 full-time Res. Assoc. employed, mainly at the Langford Vet. Station.

Ellison becomes Chairman of the UK Engineering Professors Conference and a Member of the Higher Education Quality Council.

Brett wins 3 EU contracts on 'Mechatronic tools for surgical applications', and also coordinates an EU programme on 'Minimally invasive tools for the future'.

1996   Bruce Drinkwater, Hind Saidani-Scott and Felicity Gould appointed Lects. The first becomes Sen. Lect. in 2002, Reader in

2004 and Prof. of Ultrasonics in 2007. The second becomes Sen. Lect. in 2007, and the third resigns in 2004.

Quarini appointed Editor of the J. of Process Eng. (Proc. IMechE, Part E).

For the RAE, Dept. declares its research groups to be Thermofluids, Dynamics and Control, Manufacturing and Robotics, and Materials, Structures and Design. New Univ. Research Centres have been created in the first, third and fourth of these topics. Grade 4B is obtained.

1997 Stuart Burgess (from the Eng. Design Centre, U. of Camb.) and Jeremy Burn (from the Univ. Equine Sciences Unit) appointed Lects.; the former becomes Reader in 2001, Prof. in 2005 and HoD in 2004.; the latter becomes Sen. Lect. in 2007.

Quarini is founder partner (with Imperial College and Aston Univ.) in the 'Advanced Studies in Distillation' partnership, providing CFD expertise.

Institute of Grinding Technology, jointly with Eng. Maths., sets up a TCS with RHP Aerospace in Cheltenham for 'Development of an expert system to model the grinding process'.

1998 The Univ. TQA Panel reports that 'Mech. Eng is already a successful Dept. and has much potential'. It was impressed by the high quality of students, the rigorous admissions process, the outstanding teaching laboratories, and the good tutoring system. In research it commended the close links with both Industry and with other Depts.

1999 Dept. members lead the investigation, preparation, submission and execution of the Faculty's successful BLADE proposal.

Based on CFD experience, Quarini wins a £300k MOD contract for fire-fighting on naval vessels.

2000 Christopher Truman and David Wagg appointed Lects.; the former becomes Sen. Lect. in 2005 and Reader in 2007; the latter becomes Reader in 2006.

Drinkwater obtains a 5-year EPSRC Advanced Fellowship on 'The interaction of ultrasound with solid-state interfaces'.

Following research on pipe-cleaning using the 'ice-pigging' process, Quarini creates a Univ. spin-out company CIP (Clean Ice Pig).

Stoten is the PI in an EPSRC award 'Adaptive control of systems with non-linear and chaotic dynamics with application to flight dynamics'.

2001   Morrison dies.

Stoten's Control Group develops MCS-based control and synchronisation techniques for dynamically sub-structured systems (DSS) leading to research collaboration with Japanese institutions.

Quarini obtains £720k DEFRA award for advanced food manufacturing research involving ice-pigging. Companies involved include Unilever, RHM and Geest. Income from CIP is used to support research and EngD students.

Burgess is co-author of a book *Designing Capable Reliable Structures.*

Conforming to Univ policy, research has been concentrated into three groups – Materials, Systems, and Processes. A Deptl. Research Ctte. has been established composed of the Head of each group, with an independent chairman. Its budget of £60k is to be used to support newly appointed staff.

Julian Booker and Andrew King appointed Lects.; both become Sen. Lect. in 2007.

In the RAE a Grade 5A was obtained.

2002   Burgess responsible for developing the solar array deployment mechanism used in the £1.4 billion ENVISAT satellite launched in Feb. 2002. The array is the largest and most advanced ever built in Europe. The entire mission depended on the array working within 50 minutes of launch.

Paul Wilcox appointed Lect.; becomes Sen. Lect. in 2006 and Reader in 2008.

2003   Burgess is co-author of a book *Process Selection*; a second edition was published in 2008.

Simon Neild appointed Lect. and Sen. Lect in 2008.

EPSRC 'SUPERGEN' programme awards Solid Mechanics Res. Group (with Univ. Interface Analysis Centre) £2m 'to develop an integrated toolbox to extend the life of existing UK power plants'.

2004   G. F. C. Rogers dies.

Within the BLADE reconstruction of Queen's Building, the Advanced Control and Test (ACT) laboratory is opened for the pursuance of research in the fields of Automatic Control and Dynamic Sub-Structuring (DSS).

A link with Wuhan Univ. in China is established for DSS-testing of magneto-rheological dampers for vibration suppression in bridges.

Stoten awarded Roy. Soc. Travel Scholarship for research visits to Japan.

Wagg obtains a 5-year EPSRC Fellowship for research on 'Dynamic control of hybrid systems with application to real-time dynamic substructuring'. The EPSRC also awards him £60k for 'Hybrid numerical and experimental testing for complex engineering systems'.

Drinkwater and Wilcox co-founders of an EPSRC Research Centre in Non-destructive Evaluation, having an annual budget of £1.4m. It involves 6 universities and 12 industrial organisations. It was renewed in 2007 for a further 6 years.

Smith and Edward Kingston (p.g. student) launch a spin-out company – VEQTER Ltd. – to commercialise novel residual-stress measuring techniques. In its third year of trading it had a turnover in excess of £0.5m.

Smith appointed Faculty Director of Research.

Burn et al. (the Bristol Robotics Lab.) awarded £175k for research into 'Running robots that can negotiate rough terrain'.

2005   Truman obtains a 5-year EPSRC Advanced Fellowship on 'A novel framework for predicting, measuring and analyzing weld-induced residual stresses'.

Anton Shterenlikht appointed Lect.

2006   Due to its existing nuclear energy link with British Energy, EPSRC invites the Dept. (with 5 other universities) to create an EngD 4-year degree course in Nuclear Engineering; John Bouchard is awarded a 4-year Roy. Soc. Industry Fellowship to assist with this.

Guido Herrmann appointed Lect.

Clive Wishart appointed Teaching Fellow (TF) and becomes Sen. TF in 2007, having been appointed to the Tech. Staff in 1984.

Wagg takes part in organising the Colston Research Society Symp. On 'Adaptive Structures'.

2007   Wilcox obtains a 5-year EPSRC Fellowship on 'Qualitative structural health monitoring for damage detection'.

Smith receives a Roy. Soc. Wolfson Research Merit Award.

Anthony Croxford appointed Lect.

Smith, as Director of the solid mechanics research group, links with the Safety Systems Research Centre in Civil Eng., to form the British Energy/Univ. of Bristol Research Alliance for Systems Performance Research.

Stoten's control group implement the first real-time substructuring application for aerospace engineering, involving tests on lag damper-rotor blades for Augusta-Westland.

Quarini's Clean Ice-Pig Company starts a demonstration project with the Nuclear Decommissioning Agency.

Alemzadah develops proof-of-concept technology for a jaw simulator for testing dental projects.

Drinkwater establishes a start-up company (Tribosonics) to develop ultrasonic lubricant-layer technology.

2008   Quarini receives the Univ. 'Teaching and Learning' prize for his innovative approach to the teaching of the scientific principle of engineering.

Wagg is co-editor (with O. S. Bursi) of a book *Modern Testing Techniques for Structural Systems*.

Drinkwater obtains a £4m EPSRC award, jointly with 3 other Univs., for 'Electronic sonotweezers, particle manipulation with ultrasonic arrays'. Its aim is to develop ultrasonic manipulation of small particles in fluids – with applications in medicine.

For the RAE, research is concentrated into 3 groups – Solid Mechanics, Dynamics and Control, and Design and Process Eng. In the 7-year period covered by the RAE, research funding amounted to £10.5m, mostly from the EPSRC; 6 new academic staff appointed. There have been 2 Personal Chairs, 3 Readerships and 6 Sen. Lects. In addition to the 2 spin-out companies (Veqter and Clean Ice Pig), a link has been established with Sonatest Ltd. to develop an ultrasonic wheel probe array (with EPSRC and DTI funding) for faster, non-destructive evaluation of aerospace structures.

Pavier appointed to a Personal Chair in Mechanics of Materials.

Ravi.Vardyanathan appointed Lect.

In the RAE the Dept. was placed 6th in the UK listing, from 33 Units of Assessment.

Motor Car – Automobile Engineering, 1909–35

This section has been added here because its activities were very much in what we now think of as Mechanical Engineering, and it was subsumed into that much larger department just as soon as it was convenient to do so. Reference to the 1911 entry above will show that, through the activities of F. R. B. Watson, parallel work was being carried out in Mech.Eng.

1910   Prof. William Morgan was HoD, accompanied only by a Res. Asst. The only information about him is that he was a Research Engineer at the Daimler Motor Co. in 1904, and his research record in the early days of the Faculty, was second only to that of Prof. David Robertson in Electrical Eng.

1911   The Dean's Report lists five publications by Morgan, two of the topics being 'Ignition temperature and combustion phenomena of various fuels', and 'The design of experimental paraffin carburettors'. The Colston Res. Society were funding his research, but it was also recorded that he carried out industrial work for several companies.

1912   Morgan published four papers in this session on topics such as 'Measurement of horse-power by means of fan dynamometers'.

1913   The Dept. changed its name to Automobile Eng.

1920   L. Marsden was appointed Asst. Lect. in Aeronautics (*sic*), but was not present two sessions later. Morgan's only published paper was on 'Control of gas flow in 2-stroke engines'.

1923   From this session until 1928 Morgan wrote a series of papers on the theme of 'Phenomena of combustion in internal combustion engines'.

1929   Morgan was elected President of the Institution of Automobile Engineers.

1933   It is recorded that Morgan 'was still consulting with Messrs. Lister and Co. Ltd. of Dursley, on various problems relating to airless injection internal combustion engines'. He retired in 1935 and died in 1945.

Appendix 1

A Summary of the Agreement between UCB and MVTC in 1909

1   The Society will provide and maintain in its Technical College the Faculty of Engineering in the University but it will not come into existence until the University constitutes its Faculty of Science.
Comment: This is to ensure that the University removes Engineering from its Science Faculty.

2   The Faculty of Engineering shall include civil, mechanical, electrical, mining, sanitary and motor car engineering and other subjects connected to the science of engineering as may be assigned to the Faculty.
Comment: Use of the word science indicates that training remains in the MVTC.

3   So long as (1) above maintains, the MVTC has the exclusive right to teach all students of the Faculty who are working towards university degrees, diplomas and certificates in engineering.
Comment: The MVTC controls high-level as well as lower-level teaching.

4   Funds paid to the University by any body or person for use in the Faculty must be paid to the Treasurer of the Society.
Comment: A reasonable condition given that the Society becomes financially responsible for the Faculty.

5   All engineering plant and equipment belonging to UCB which is fit for purpose must be transferred to MVTC.

6   The Society to provide in the Unity Street building all facilities necessary for the educational requirements of the Faculty, but will be able to reserve such facilities as are required for the MVTC Secondary School and Evening Classes, provided they do not interfere with the essential needs of the Faculty.
        Comment: In fact, the three levels of education worked harmoniously together from the very beginning.

7   If the University were to decide that the arrangements made by the Society for accommodating the Faculty were not satisfactory, a Board of Arbitration would be established to consider the issue, consisting of three people, the Lord President of the Privy Council and one person each nominated by the Presidents of the Institutions of Civil and Mechanical Engineers. Any recommendations of this Board would either be accepted by the Society and the required actions carried out, or it would give up the Faculty.
        Comment: The Board never met.

8   If the Society ever decided to give up the Faculty, all facilities originally transferred from UCB would be returned to the University but at a rate which would not inconvenience either party.

9   Courses of study in the Faculty are to be determined by the University and none of its academic staff can be required to work in the MVTC without the approval of the University. This does not apply to MVTC staff already in post.
        Comment: To be doubly sure, the Agreement states that the MVTC shall have full control of non-Faculty teaching activities.

10   The degrees of B.Sc., M.Sc., and D.Sc. shall be open to all students in the Faculty, *expressly including* (my italics) students of the Evening Classes. There will also be a University Certificate in Engineering for those not having matriculation.

11   Although fee levels shall be determined by the University, the Society reserves the right to remit up to 60% to students who cannot afford to pay.

12   The seven academic staff then employed by the MVTC shall be transferred to the Faculty under existing terms which includes the power of dismissal by the Society alone. For the four academic staff transferred from UCB to the Faculty, dismissal shall be considered separately by the University and the Society; any disagreement to be referred to the Privy Council.
Comment: Wertheimer succeeded in making only one of the UBC personnel redundant.

13   The appointment (and dismissal) of the Principal of MVTC shall rest exclusively with the Society. The present Principal shall be a Professor in the Faculty, but the status of future Principals would rest with the University.

14   The PRESENT Principal of MVTC and every succeeding Principal shall be Dean of the Faculty.
Comment: The word 'present' appeared in capital letters in the actual Agreement, more than likely at the insistence of Wertheimer who occupied the post until his death in 1924. The only other Principal to become Dean was Andrew Robertson who was so outstanding as both researcher and administrator that no difficulty arose.

15   The Society could nominate ten persons to the University Court and three to its Council.

It is to be noted that the relationship between the University and the MVTC matured greatly from this time on, with members of the Society taking a prominent role in the affairs of both the Faculty and the University as a whole, and even after its responsibility ended in 1950, the Society's contribution continued at the same high level through the knowledge and experience of those of its members who served on the Council of the University, several as its Chairman, as well as on committees for the appointment of senior staff, and for the construction of new buildings. Fig. 4.4 shows the Chancellor, Winston Churchill, unveiling the Society's commemorative plaque in the entrance hall of Queen's Building.

Appendix 2

The Faculty and the Engineering Institutions

A2.1 Introduction and Early History

No account of engineering education in the UK can be complete unless it considers the unique influence exercised by the Engineering Institutions; they perform a significant role in the process of qualifying engineers for professional status, a role which is performed in all other countries either by state or central government. Just as universities in the UK are autonomous, charitable, bodies, so are the Institutions, and they continue to be free from any form of Government control, although as we shall see, the Finniston Report in 1979 did suggest that statutory control should be introduced.

To understand how this unique situation has come about, it is necessary to record that in the year 1818 eight young engineering apprentices in London were learning their profession by copying the activities of their seniors – and paying for the privilege of doing so! They came to realise that they were being trained in the accepted design and construction procedures of the time, rather than educated in principles, and, regarding this as unsatisfactory for their future progress, they formed themselves into a coffee-house club, meeting regularly to discuss problems which they were facing in their work. They were soon joined by others in formalising their meetings to create a society which they referred to as an Institution of Civil Engineers (ICE), with 15 founding members. Their expressed aim was 'for facilitating the acquirement of knowledge in the civil engineering profession', and to this end they asked the profession's most eminent exponent, Thomas Telford, to be their President. He did so, and occupied this post until his death in 1834. Telford himself had received no formal education, but became the supreme builder of roads, bridges, churches and canals in England, Sweden and his home country Scotland, and, as a reflection of his status, is one of only two engineers to be buried in Westminster Abbey. Robert Stephenson is the other, and although he is remembered principally as the builder of railways, he regarded himself as a civil engineer and, in fact, became President of the ICE in 1856. Both Telford and Stephenson were elected to the Royal Society.

We make the important note in passing, that the term 'civil engineering' had been deliberately chosen to encompass at this time every aspect of engineering which was deemed not to be of a 'military' nature, formal education and training for which had been established in France by its government as early as 1720 for officers of the Corps of Engineers of Ways of Communication, which led in 1747 to the creation of the now prestigious *Ecole des Ponts et Chaussées.*

Without Telford's influence in political and royal circles, it is doubtful if the ICE would have survived beyond its first few years, but it did so, having, in due course, a major influence on engineering education. His significant role in the development of Scotland's infrastructure drew him to the attention of Queen Victoria, which makes it less than surprising that in 1840 she established a Regius Chair of Civil Engineering and Mechanics at the University of Glasgow, the first University Chair of Engineering in the UK. Its second occupant was William McQuorn Rankine, a name which students of today are familiar with, because his outstanding achievements in research on such subjects as elasticity, soil mechanics, thermodynamics, ships' propulsion and electricity are still part of their undergraduate studies.

In its early years the ICE's growth and prestige was a direct result of the development, both at home and worldwide, of canals and then railways, both of which in the UK required Acts of Parliament involving the expert advice of ICE members such as Telford. It became necessary therefore for these engineers to establish their offices as close as possible to Parliament, and to create a meeting place for discussion and preparation of the documents which they were required to present. This became the ICE's first headquarters – in what is now H.M. Treasury, not far from its present position in Great George Street.

A2.2 The Institutions and University Education

As the scope of engineering developed during the second half of the nineteenth century, the ICE was not able to satisfy all the specialist needs of those involved, so that groups broke away to form similar learned society bodies of their own. The first group to do so was led by George Stephenson, father of Robert, to found the Institution of Mechanical Engineers (IMechE) in 1847, followed by the Institution of Electrical Engineers (IEE) in 1871. Rather surprisingly, the Royal Aeronautical Society (RAeroSoc) was founded as early as 1866 by the Duke of Argyll who had a passionate interest in developing heavier-than-air machines, some 34 years before the Wright brothers succeeded in doing so. As time progressed, the major Institutions were themselves subjected to breakaway groups which formed smaller Institutions dedicated to specialist interests. For example, the Institution of Structural Engineers formed a subset of civil engineers, and the Institution of Production Engineers a sub-set of mechanical engineers. The Institution of Chemical Engineers (IChemE), founded in 1922, can also be regarded as developing from the IMechE. From our own Faculty viewpoint, because the Computer Science Department became part of it in 1989, mention should be made of the British Computer Society which was created in 1957.

In the beginning, the Institutions were learned societies, concerned with acting as meeting points for their members to discuss engineering issues and, through the formation of libraries, as repositories of professional knowledge in the form of books, drawings, and construction and manufacturing details. But it was not long before they obtained Royal Charters which allowed them to engage in a second prime function, that of 'Qualifying Bodies', that is to say, organisations which, through examinations of various types, accepted candidates into membership and conferred professional titles. There can be conjecture about their reasons for taking on this role, but it is likely to have been that senior members, having built up a body of knowledge and experience in successfully carrying out innovatory and complex projects, wished to ensure, by examinations which they introduced, that their successors were capable of emulating them for the public good. There were at that time no university courses in engineering which they could use for the purpose, requiring them to develop their own, together with viva voce examinations and assessments of professional experience.

As soon as university courses became available, the Institutions took advantage of them, in many instances through their senior members being invited to help construct such courses; clearly there was mutual benefit in this. In Bristol, before the Engineering Faculty was created, engineering students at UCB and MVTC were entered for Institution examinations and we recall that MVTC regarded itself as superior to UCB because of its greater success in these examinations. But in the very first year of the Faculty (1909), it appointed to its board Sir James Inglis; not only was he the Manager of the Great Western Railway but was also President of the Institution of Civil Engineers. The remit given to him by the Faculty Board was '. . . to shape the content and style of the course so that it related to the real engineering world'. Through such appointments in the various universities, the Institutions were able to assess the value of courses of instruction, and where appropriate, to give exemption to their students from the Institution's own examinations.

As a general rule the Institutions were not prescriptive regarding the actual course content, preferring that different universities should have strengths in different areas, but from time-to-time they have requested the inclusion of certain subjects. An illustration here occurred in the 1960s, when instruction in Law, Politics and Management were introduced as part of the 'Engineer in Society' courses.

The passage of time has seen great changes in the educational requirements of different Institutions and this has prompted widening rules for membership. For example, although relevant science degrees have always been accepted by the IEE and the RAeroSoc – subject to satisfactory post-degree practical training and experience – it was only in the 1960s that the ICE began to accept degrees in such subjects as physics, mathematics, geography, geology and economics, indicating an acceptance of the widening requirements of the civil engineering profession.

Because of their dominance in size of membership, it is convenient to describe the ICE, IMechE and IEE collectively as the 'big three' and their interfaces with the Faculty have many common characteristics. It was usual, for instance, for the Local Associations of these Institutions to hold their meetings in Queen's Building, a habit which served three purposes. It allowed students to become familiar with professional requirements; it allowed students and academic staff to meet with practising engineers, and it allowed this last group an opportunity to assess the quality of both staff and students. All these opportunities were of importance when the Institutions came to assess the quality of the courses, which is still that of a two-day visit by a group of senior members who investigate the quality of students, staff, teaching and research. Of course, the last two of these topics are now also assessed by bodies appointed by the Higher Education Funding Council but for a different purpose, that of allocating Government funding.

From the 1970s onwards the number of universities offering degrees in engineering increased appreciably, so much so that the 'big three' Institutions became concerned at the low academic level which a number of them appeared willing to accept for their students, even though the output levels – as measured by the degrees awarded – at these same universities were not noticeably dissimilar from those at the older established universities. It can be argued of course that the correlation between input and output is uncertain, but the 'big three' were sufficiently concerned to set up their own individual studies of the situation (Ref. 13), before setting down specific intake standards based upon A-level (or Scottish Higher Certificate) performance, either of individuals or the average of the intake cohort. The Faculty never had any difficulty in meeting the rather modest standard requirement imposed by the 'big three', of three C-grades at A-level for each student.

A2.3 The Finniston Committee and the Engineering Council

By the 1950s the proliferation of Institutions awarding professional engineering qualifications was accepted as one contribution to the lack of esteem accorded to the UK engineering profession by the public at large. But of greater importance was the message from Government Departments that they had difficulty in knowing which of these Institutions carried the necessary authority to be consulted in the process of shaping Government policy. Sensitive to this, the 'big three' Institutions attempted some rationalisation by creating the Engineers Joint Council, but this lasted only four years, and it was not until 1962 that a further attempt was made, resulting in the establishment of an organisation which grew into the Council of Engineering Institutions (CEI), a body which received a Royal Charter in 1965. This body made progress in enhancing the status of UK engineering in two areas. First it set in motion the creation of an elite body – to be called the Fellowship of Engineering, membership of which was to be by peer-group election. It was established in 1976, received a Royal Charter in

1983 and became the Royal Academy of Engineering (RAE) in 1992. The second achievement of the CEI was the categorisation of engineers into three groups according to length and level of education and training. At the top were Chartered Engineers (CEng); but to give status to the many non-graduates who performed essential engineering functions, but had been recognised in an unsatisfactory manner by some of the Institutions, two categories were introduced, those of Technician Engineer (TEng) and Engineering Technician (EngTech). At a later date the first of these became Incorporated Engineer (IEng) in order to indicate that they occupied a middle role between the engineer and the technician. Inevitably, the standards set by the CEI for the CEng category were compromises which the major Institutions regarded as being too low, and therefore refused to accept. Instead, as discussed earlier the 'big-three' reviewed their own requirements, proposing new ones where necessary.

The failure of the CEI was due largely to this failure to reconcile the greatly differing standards required by the many Institutions, each jealous of its own autonomy as a qualifying body and unwilling to become simply a learned society. Such a situation, in one of the UK's most important industries, was bound to attract the attention of the Labour government of James Callaghan which was concerned with European collaboration and competition, and it came in July 1971 through the establishment of 'A Committee of Enquiry into the Engineering Profession' (Ref. 4) to be chaired by Sir Monty Finniston with the following terms of reference:

To review for manufacturing industry and in the light of national economic needs

(1) the requirements of British industry for professional and technical engineers, the extent to which these needs are being met, and the use made of engineers by industry;

(2) the role of the Engineering Institutions in relation to the education and qualification of engineers at professional and technical level;

(3) the advantages and disadvantages of statutory registration and licensing of engineers in the UK ;

(4) The arrangement in other major industrial countries, particularly the EC.

The above terms of reference have been given in full because – perhaps significantly – membership of the Finniston Committee was noticeably lacking in high-level representation from the Institutions. This was important, because if its recommendations were to be accepted and implemented by Government, it is possible that they would make a serious change to the existing interface between universities and the Institutions.

Of the many recommendations made by Finniston in late 1979, the most significant was that an 'Engine for Change' (Ref. 20) should be established by Government to be described as the 'Engineering Authority'. It is certainly significant that the Government which received the report, and acted upon it, was the Conservative one of Margaret Thatcher, and instead of a statutory body, an 'Engineering Council' (EngC) with charitable status was established in 1981 with 53 Institutions in membership. Although some Government financial support was provided, other income was derived from its role in maintaining a register of the three categories of engineers established by the CEI. Because this income was collected for it by the group of Institutions licensed to award these qualifications, it was they who retained the real authority.

From the beginning, the relationships between the Institutions and the EngC were not auspicious, partly because the Institutions were not given representation on the Council, and partly because, by establishing its own examination system, it was seen as an unnecessary additional participant in the assessment of professional engineering qualifications; moreover, it was unlikely to add anything to the learned society role. There was also a controversial early statement from the EngC that it, unlike the Institutions, was concerned in the greater task of promoting industry and commerce in the UK for the public benefit, whereas – the EngC said – the Institutions were not concerned with national issues, but took their primary duty to be that of their own disciplines and memberships.

The Institutions did however see the value of the EngC in establishing common standards for the three categories of engineers, CEng, IEng and EngTech, through its 'Standards and Routes to Registration' (SARTOR) documents and their consolidation into a computer-based register.

In practice, the links between the Institutions and the universities continued much as before, with the major Institutions being 'licensed' by the EngC to admit to the CEng category through a four-year MEng degree. By 1996, the influence of the 'big three' persuaded the EngC to recommend that only universities having at least 80 per cent of its engineering students with 3B-grades at A-level (or their equivalent) could award the MEng degree. The Faculty had no difficulty in meeting this requirement, and together with several of the other prestigious universities, it had offered all its students the option of the 4-year MEng degree. This was partly due to the basic science preparation at A-level having deteriorated in quality to the extent of requiring additional remedial courses at university if satisfactory standards were to be maintained, and partly because, with membership of the European Community in mind, they wished to give students the opportunity of studying for one year in a European university.

By the mid-1990s the Institutions became increasingly exasperated with the EngC and instigated a review of its function, resulting by late 1996 in a change to its charter, which meant that its appointed council was replaced by a senate, the majority of whose members were now elected by a process dominated by the Institutions. This change was not a success, largely because the senate was an unwieldy body of 54 members. A little later, the Institutions launched an 'Activity Review' of the EngC, the purpose being to ensure that it became responsible only for those activities which were common to all the Institutions. However, this resulted in a number of hitherto successful activities being terminated or hived off, thereby weakening the EngC's role.

In 1999, at the prompting of Tony Blair's New Labour Government, a new initiative was launched, which resulted in 2002 in the division of the EngC into two parts, an Engineering and Technology Board having the original 'Engine for Change' role, and a new charter for an Engineering Council UK (EngCUK) concerned with regulation of the profession and its representation internationally.

Appendix 3

The Faculty and the School for Advanced Urban Studies (SAUS)

A3.1 Introduction

As with the creation of the Department of Architecture, it was the Civil Engineering members of the Faculty who played a part – though considerably less significant than with Architecture – in the creation of SAUS. Their motive here being the expectation of development in transport studies, an important sub-discipline of civil engineering which at that time received a very modest level of attention in the Faculty. They had some justification for this expectation, because the Government report presented by Lady Sharp in October 1970 proposed that 'A Centre should be set up for the advanced training and education of urban *transport planners* (my italics) . . . if possible to cover land use planning as well.' As we shall see, a centre was set up in Bristol, but the sequence of events which followed present a cautionary tale for universities which respond to initiatives promulgated by officials in Government who are not able, for a variety of well-known reasons, to pursue their ideas to a successful conclusion. Here, one such reason was the replacement in June 1970 of the Labour government of Harold Wilson by the Conservative one of Edward Heath. For SAUS even before it was created, the intentions of Lady Sharp regarding its mission had been turned away from transport studies towards more general aspects of urban planning, and of probably greater importance was the fact that the University had no role in the appointment of its first Director.

A3.2 **The Government's Report on Transport Planning**

The spur for the University's involvement came from Gordon Mills, a member of the Department of Politics who had inside knowledge of the Government's intentions, which he communicated to Prof. Ashworth, the Dean of Social Sciences. By March 1970 the details of the Government's intentions were known, and included the creation within a university – as yet unspecified – of a 'Centre for Teaching and Research in Transport Planning'; it was this repeated reference to transport planning which excited the Faculty's interest. Through personal contacts in both the UGC and the Ministry of Transport, Ashworth had discovered that Bristol was one of the universities to be invited to bid, prompting him to call an informal meeting of interested colleagues. Mills wrote a discussion paper for the meeting, which showed evidence of detailed information about the aims and objectives of the proposed school, in particular that a two-year Masters degree was envisaged based mainly on course work. Regarding finance, the capital costs of a new building up to £500,000 would be provided together with running costs for the first five years. Intended students would be those whose aim was to work in planning in local and central government. Shorter, post-experience courses would also be provided.

In attending this meeting, the Civil Engineering representatives were aware that the Science Research Council would shortly establish five University Research Centres in Transport Studies which would be totally committed to the engineering aspects of transport – and that Bristol was not one of them. But they took the view that the proposed centre in Bristol needed an awareness of the engineering constraints on what it was possible to achieve, and this was accepted by the Ashworth meeting.

The essential outcome of the Ashworth initiative was to suggest to the Vice-Chancellor, Alec Merrison, that a working party be established to produce detailed arguments for Bristol as the location for the centre. Simultaneously, the Bursar was informed that were Bristol to be successful, a combined teaching and residential building would be required for 100 mature students by the autumn of 1972.

Bristol's case was set out in a paper described as the 'Ingredients For Training In Urban Planning' written by Ashworth and Michael Chisholm, a Professor in the Geography Department. Although the word 'transport' had been omitted from the title, the paper contained very specific details of what the ingredients of the course would be. Of its five sections, the first, described as 'Basic Analytical Techniques and Concepts' contained such topics as linear programming and network analysis, statistics and data management, whilst the second, within the title 'Technological Considerations', included traffic engineering and environmental engineering, together with an explanatory note that

> This section is intended to give the planner some acquaintance
> with problems and techniques in these areas, and to enable
> him to communicate with *engineers and others* (my italics)
> who may carry the principal responsibility for the
> consideration of these matters.

A3.3 The University's Bid

Ashworth was given to writing long but informative and well-constructed letters, and in May 1970 one such suggested to the Vice-Chancellor that the University should put the discussions on a formal basis, operating through the established University bodies. It is not known whether at this stage the Vice-Chancellor was supportive of the bid being made, but his brief reply to Bill Ashworth possibly offers a clue, because he suggested referring the matter to the Long Term Policy Committee! As sometimes happens in a university, organisational lines became crossed, in this case by the reports of the Committees of Professors in Social Sciences and Engineering on this issue being discussed prematurely in Senate, the result of which was the establishment of a Senate Committee on the proposed Centre. Its first meeting took place in July 1970, and although the official invitation to apply had not been received, it was expected during the Long Vacation, with a reply required by October. It was agreed that the 'Ashworth' working party would draft the reply for approval by the Committee.

In the meantime, the Bursar had been active in referring the matter to the Developments Committee who recommended that a site should be provided. Although he had retired in 1968, Sir Alfred Pugsley continued as a member of this Committee, and knowing that the modifications to Rodney Lodge had been made so that it could be extended at a later date, he wondered whether such an extension might provide the necessary accommodation.

The invitation to bid was not received until October 1970, with response required by the end of the year. It had been extensively revised in comparison with the original, bearing clear signs of attempts to achieve the desired objectives at the cheapest rates. It has to be assumed that the Bristol bid was made in due time, but reference to it does not appear again until October 1971 in an exchange of letters between Ashworth and the Vice-Chancellor referring to a visit to Bristol by four senior officials from the Department of the Environment. There was of course changes in Government at that time, resulting in delays to new initiatives. It is not clear if the decision had already been made to award the Centre to Bristol, but Common Room gossip was that Sir Colin Buchanan, author of the earlier Government report on 'Traffic in Towns', and recently Adviser to the Ministry of Transport, had already been designated as the first Director of the Centre, and had been given the choice between Bristol and Oxford. It subsequently transpired that the first, but not the second, component of this gossip was correct. At that stage the DOE party had five possible universities on their list, later reduced to two – Bristol and Southampton.

The DOE officials met with a group of University members in November 1971 in Bristol to discuss a detailed agenda. That there was not complete unanimity of view between the two groups can be gauged by the Vice-Chancellor stating that because there would be so many departments contributing, the Centre must be an integral part of the University; from the other side, the DOE leader (J. D. Jones) said that the Centre would be dealing with '. . . family-based men who would want comfort and privacy and would not want to get mixed up with the student body.' With hindsight it can be seen that here was a polarity of view about the role which the University would be expected to play, a polarity which would quite soon lead to the lack of any serious level of involvement by University academic staff in the affairs of the Centre.

Notwithstanding the above differences, the leader of the DOE group (now Sir James Jones), was accompanied in January 1973 on a second visit to Bristol only by Sir Colin Buchanan, at which it became clear that Bristol was the favoured site. It was not known whether this was due to the fact that Buchanan then lived in the Cotswolds, making his travelling arrangements easier than to Southampton, or that their bid, it was said, had been put together principally by the Engineering Faculty there.

The offer to Bristol to house the School for Advanced Urban Studies was accepted by the University Council in June 1972, with a formal announcement by the Secretary of State a few days later, indicating that its scope had been widened to cover the urban environment generally, and was not now restricted to transportation and land-use planning – a fact which made it less likely that the Faculty would have a significant involvement. To nobody's surprise, but not to everyone's satisfaction, particularly as the University had not been consulted, the first Director was to be Sir Colin Buchanan.

A3.4 The Early Years

The first meeting of the Committee of Management of SAUS took place in February 1973 with the Vice-Chancellor as chairman. Its members consisted of Ashworth, Buchanan, Dineley (Geology), Jones (Architecture) and Severn (Civil Engineering), together with Lady Sharp, the Chairman of Council (Richard Hill) and two members of Council. Buchanan was formerly appointed Director and C. A. North the secretary. A Board of Studies was established, and a prestigious site next to Wills Hall in Stoke Bishop offered for the new building, with temporary accommodation in Woodland Road and possibly Rodney Lodge.

In the important matter of the purpose of SAUS, Lady Sharp continued to press for two-year postgraduate courses as its main function, but the academic members (which included Buchanan as a part-time Professor at Imperial College), having had the dismal experience of trying to recruit students for even one-year MSc courses, insisted that shorter post-experience courses must be offered.

At the next meeting of the Committee, Buchanan presented for discussion an 11-page collection of his thoughts on 'Management of Urban Change', being his blueprint for the future of SAUS. Even now, 30 years later, it appears as an impressive document, which if carried through would surely have created a successful SAUS. Surprisingly however, he proposed that instead of starting from cold, SAUS should develop from his own commercial creation – the Planning and Transport Research and Transportation Company (PTRC), which would move from Imperial College to Bristol. For a Professor to be the principal director of a commercial enterprise was clearly acceptable to Imperial College, but not to Bristol at that time, and the decision not to accept Buchanan's proposal may have precipitated the announcement of his intention to resign during the summer of 1975.

Early in 1975 it became known that the City had refused outline planning consent for the Stoke Bishop site, and rather than appealing against this decision, there was a general wish to switch attention to an enlargement of the Rodney Lodge site. After the DOE approved this action, Powell and Moya were appointed architects in June 1976. Following the resignation of Buchanan, P. A. Eddison – an early appointment to SAUS in 1975 – became the Director, with Murray Stewart as his deputy.

A3.5 Finance and the Teaching Programme

The agreement between the University and the DOE was that building costs would be fully covered by them, as well as recurrent costs for the first five years, which meant until August 1979, after which the University would be responsible. But the understanding between SAUS and the University was that whatever activities in teaching and research they choose to engage in, they should be self-financing by the end of the 1978 session. From the SAUS viewpoint this constrained their academic ambitions and caused them to search around for almost any organisation which could afford to pay for short courses, often on topics far removed from the original SAUS concept. For examples, it began to encroach on the interests of other departments in the University by offering courses on such topics as racial equality, and on introductory courses for newly-elected members of both the House of Commons and the European Parliament. Only those to the former body were actually carried out, and then only for a short period. When it proposed a new one-year MSc course on Public Policy Studies, the Vice-Chancellor was caused to remark that '. . . the third letter of the SAUS acronym was 'Urban' and this was missing from the title of the proposed course'. Nevertheless, it was allowed to proceed despite obvious conflicts with courses already provided by the Faculty of Social Sciences and Continuing Education. All these Departments required an upper second-class first degree for admittance to its MSc courses, but the SAUS management was not prepared to accept this, and successfully argued for more flexible entrance requirements in view of the type of applicant it was hoping to attract.

A3.6 The Effect of the 1981 Financial Cuts

The University's self-financing understanding with SAUS meant that its financial performance was carefully monitored. It was with some surprise therefore that in the session 1975 its deficit was £148,000. At a meeting of the SAUS Management Committee in November 1977 the Vice-Chancellor said that he could not see how SAUS could ever be self-financing and that the University must discuss its long-term future. It was given a short term relief by the DOE agreeing to fund the deficit for a further year, until August 1980. However, the 1977 deficit had already risen to £178,000, causing the University to establish a committee to look into its long-term financing.

At the November 1980 meeting of the SAUS Management Committee it was announced that the Rodney Lodge building programme had been completed at a cost of £848,700, and had been officially opened by Tom King – Minister of State for Local Government (and brother of Stella Clarke, a future Chairman of the University Council). The building programme had certainly had its effect on the ability of SAUS to carry out its teaching programme as planned, resulting in a deficit for 1980 of £152,000 with a similar deficit forecast for 1981. But in November 1981 the UGC cuts of 15 per cent in the University's finances (Chap. 6) meant that SAUS, as a unit for which the University was financially responsible, but which had made itself largely detached from mainstream activities, found itself in a very exposed position. It is likely that it was saved from closure in the Vice-Chancellor's proposed solution to these cuts by the fact that the University was contractually obliged to the DOE to use Rodney Lodge for 20 years for SAUS-type activities. But as its response to these financial constraints, SAUS was required to make a cut in expenditure of £58,000 p.a. by 1984 through an increase in its self-financing activities from 64 to 78 per cent; to formally associate itself with the Faculty of Social Sciences, and to reduce the number of full-time staff. It was agreed that existing full-time staff would face the risk of redundancy if these measures were not implemented.

The achievement of these financial objectives fell to Murray Stewart in September 1982 following Eddison's resignation as Director for personal reasons unconnected with the financial difficulties. By a range of measures, which included sub-letting facilities in Rodney Lodge and developing courses in information technology and on housing benefit for the DHSS, he had made some progress by April 1983 in achieving the financial targets. Even so, in June 1984 the University asked SAUS to make further reductions in its financial demands and established a Working Party of the Finance Committee to consider its longer term financing. One suggestion made was that SAUS should be integrated with the School for Applied Social Studies (SASS) and fully absorbed into the Faculty of Social Sciences, but this was not achieved until 1987.

Murray Stewart's aim was to transmute SAUS into what he referred to as 'A National Centre for Teaching and Research in Public Policy Studies', and although his financial difficulties had not been entirely resolved, he succeeded in 1988 in persuading the University to appoint Julian Le Grand as professor in this subject. In this year also, the fact that SAUS had been fully integrated into the Faculty of Social Sciences was indicated by the election of Murray Stewart as its Dean, requiring that Randall Smith become Director of SAUS for a three-year period. In 1991 Le Grand, as the new Director, succeeded in establishing full integration with the School for Applied Social Studies, particularly through his MSc course in public policy studies, and by the creation of two new MSc courses in international policy and in the economics of public policy. Some few years later SASS became The School for Policy Studies, but it too was shown by the Resource Allocation Mechanism to be living beyond its means, and in 1997 was requested to reduce staff or face compulsory redundancy.

Appendix 4

Student Numbers in the Faculty

Except for Table A4.12, the statistics of student numbers is given for 5-year intervals. The date given is for the session starting in that year. Until 1936 only the number of registered students was recorded, without division into the three departments of Civil, Mechanical and Electrical Engineering.

Table A4.1 Undergraduate students in the Faculty 1911-31

	1911	1916	1921	1926	1931
Day Students	74	83	209	86	97
% Matriculated	72	77	90	92	97
Evening Students	441	314	650	0	0

Note
In 1916, 53 of the day students were serving with the armed services.

Table A4.2 Undergraduate students in the Faculty, 1936, 1941 and 1946

	1st Year	2nd Year	3rd Year Ord	3rd Year Hons	4th Year Hons	Total
Civil		7 15 24	5 5 13	2 9 7	4 0 2	18 29 46
Mechanical	20 21 40	6 13 13	6 1 6	3 4 6	2 0 8	37 39 73
Electrical		8 11 15	5 1 6	0 5 7	2 0 3	15 17 31
Aeronautical		11	2	2		15
Total	20 21 40	21 39 63	16 7 27	5 18 22	8 0 13	70 85 165

Notes

1. In each of the main columns, figures are given for the sessions 1936, 1941 and 1946 reading from left to right.

2. The Aeronautical Engineering Dept. was created in 1945.

3. The 4th Year of the course was discontinued during the Second World War.

4. In the 1st Year students were not attributed to Departments; they are listed here under Mechanical.

Table A4.3 Undergraduate students in the Faculty for the session 1951

	Civil	Mech.	Elect.	Aero.	Total
Inter	4	3	4	6	17
1st Stage	24	18	15	16	73
2nd A	3	7	1	4	15
2nd Band C	18	13	11	10	52
3rd Hons.	7	9	8	6	30
3rd Ord.	8	6	8	6	28
Total	60	53	43	42	215

Notes

1. In this session a new format was used for presentation of data.

2. There were 7 'Higher Degree' students without attribution to Depts.

3. In 1950 there were 3 women students, 2 in Civil and 1 in Elect.

Table A4.4 Undergraduate students in the Faculty for the sessions 1956, 1961 and 1966

	Civil			*Mech.*			*Elect.*			*Aero.*			*Total*		
Inter	2	0	0	1	0	0	6	0	0	6	2	0	15	2	0
1st Stage	28	30	30	29	32	40	23	32	32	29	30	26	109	124	128
2nd A	4	8	2	4	11	18	9	8	6	5	5	6	22	32	42
2nd B+C	25	19	22	17	18	26	16	21	17	21	18	18	79	76	83
3rd A	6	7	2	6	7	1	4	8	5	4	6	2	20	28	20
3rd B	8	13	12	8	12	20	9	9	12	9	8	13	34	42	57
3rd C	7	8	8	3	9	6	5	10	4	6	17	3	21	44	21
Total	80	85	86	68	89	121	72	88	76	80	86	68	300	348	351

Notes

1. As in Table 4.2, the entries for the three sessions are to be read from left to right in each main column.

2. There were 14 'higher degree' students in the first of these sessions and 16 in the second without attribution to Depts. In the third there were 34, distributed as 8 in Civil, 7 each in Mech. and Elect., 6 in Aero. and 6 students in the MSc course 'Fluid and Thermal Studies for Industry'.

3. From 1966 (until its closure in 1985) the Faculty, notably Civil and Theor. Mech. had substantial teaching obligations in the Dept. of Architecture. In 1966 there were 93 students for the BA in Architecture, 46 for the Diploma in Architecture, and 9 for a higher degree.

Table A4.5 Undergraduate students in the Faculty for the sessions 1971, 1976, and 1981

	Civil			Mech.			Elect.			Aero.			Eng. Math.			Total		
1st Stage	34	43	33	41	46	38	35	46	41	28	30	37	0	0	17	138	165	184
2nd A/B	28	32	32	38	36	42	22	50	50	19	31	39	0	0	15	107	149	155
3rd A	4	7	5	6	12	5	1	11	9	6	6	5	0	0	0	17	36	25
3rd B	12	17	25	22	15	27	11	20	19	6	5	8	0	0	13	51	57	80
3rd C	13	11	6	17	12	11	7	14	11	11	6	9	0	0	0	48	43	36
Total	91	110	101	124	121	123	76	141	130	70	78	98	0	0	45	361	450	497
High Deg.	11	8	5	13	11	6	7	8	13	8	3	5	1	3	6	40	33	35

Notes

1. As in Table 4.2 the entries for the three sessions are to be read from left to right in each main column.

2. Of the 497 undergraduates in 1981, there were 16 from overseas, mostly in Civil and Aero.

3. Theoretical Mechanics changed its name to Engineering Mathematics in 1973 and by 1981 was offering its own undergraduate degree. In 1976 there were 428 men, but only 22 women students who were not attributed to Depts. By the same year the course structure in Architecture had changed; there were now 116 students in the BA(Arch) and 56 for the Diploma in Arch.

Table A4.6 Students in the Faculty for the session 1986

	Civil	Mech.	Elect.	Aero.	Eng. Math.	Com. Sys.	Total
1st Year	41	54	66	44	14	16	235
2nd Year	44	44	41	33	12	19	193
3rd Year	33	41	38	36	10	12	170
4th Year		8					8
Total	118	147	145	113	36	47	606
High Deg.	14+7	17+6	19+12	9+2	7+0	5+5	71+32

Notes

1. Computer Systems (Com. Sys.) was introduced in 1984 jointly by Electronic Eng., Eng. Math. and Computer Sci.

2. Only Mech. offered the 4-year course at this time.

3. In the higher degrees row the second figure is for overseas students.

Table A4.7 Students in the Faculty for the sessions 1991 and 1996

	Home u.g.		*O'seas u.g.*		*Home p.g.r*		*O'seas p.g.r*	
Civil	118	152	9	9	6	15	8	5
Mech.	206	263	5	5	12	27	2	8
Elect.	178	162	18	16	28	25	10	23
Aero.	162	218	11	13	13	15	4	2
Eng. Math.	28	51	3	0	6	12	1	2
Comp. Sci.	73	186	5	9	9	18	2	10
Total	765	1032	51	81	74	112	27	50

Notes

1. As in Table A4.2 the entries for the two sessions are to be read from left to right in each main column. In the last two columns, p.g.r. indicates postgraduate research students only.

2. From 1989 there was a wide range of courses available and the Dean's Report to Senate gives a detailed breakdown. At this time the first four Depts. offered a 4-year course in which the third year was spent in an approved university in mainland Europe.

3. Computer Science (Comp. Sci.) joined the Faculty in 1989.

4. In 1996 the Faculty overshot its MASN target by 41 students, for which the University received fees only; but in the following year the target was increased in order to support the 4-year courses.

Table A4.8 Students in the Faculty for the sessions 2001 and 2006

	Home u.g.		*O'seas u.g.*		*Home p.g.r/p.g.t*		*O'seas p.g.r/p.g.t*	
Civil	168	210	38	26	11/3	16/0	8/3	12/11
Mech	238	289	19	40	20/7	21/6	14/6	4/7
Elect.	215	154	54	86	34/25	35/17	27/18	33/26
Aero.	301	249	23	33	23/155	26/211	8/3	12/11
Eng. Math.	78	161	0	13	9/0	12/6	4/0	0/0
Comp. Sci.	283	258	29	15	29/42	32/48	6/45	11/55
Total	1283	1321	163	213	126/232	142/288	69/74	80/99

Notes

1. As in Table A4.2 the entries for the two sessions are to be read from left to right in each main column.

2. In the last four columns entries are for both research and taught postgraduates, the latter being dominated by the part-time Aero. IGDS programme, and by Computer Science.

Table A4.9 Male and female undergraduate student numbers for the three sessions 1997, 2002 and 2007

| | Home | | | | | | Overseas | | | | | |
	Male			Female			Male			Female		
Aero.	253	282	198	18	34	22	16	20	26	3	3	7
Civil	122	156	207	41	38	45	17	36	24	3	8	9
Comp. Sci.	203	269	198	22	31	14	10	33	10	6	4	0
E + E	152	186	133	19	15	16	34	54	74	12	25	25
Eng. Math.	46	64	139	12	26	52	1	1	9	0	0	4
Mech.	224	206	255	33	26	52	11	17	61	1	5	7
TOTAL	1000	1163	1130	145	173	169	89	161	204	25	45	52

Note

For the Tables A4.9, A4.10 and A4.11; in each main column figures are given for the sessions 1997, 2002 and 2007, reading from left to right.

Table A4.10 Male and female postgraduate research numbers for sessions 1997, 2002 and 2007

| | Home | | | | | | Overseas | | | | | |
	Male			Female			Male			Female		
Aero.	16	21	37	1	2	6	4	4	10	0	2	3
Civil	12	4	26	2	3	12	5	17	17	1	2	1
Comp. Sci.	14	33	32	4	7	2	7	5	13	3	3	2
E + E	27	38	34	4	2	3	20	18	24	2	5	13
Eng. Math.	10	11	15	3	1	1	0	5	0	1	0	1
Mech.	13	15	25	6	1	1	10	4	0	4	2	4
TOTAL	92	122	169	20	16	25	46	53	73	11	14	24

Table A4.11 Male and female taught postgraduate numbers for sessions 1997, 2002 and 2007

| | *Home* | | | | | | *Overseas* | | | | | |
	Male			*Female*			*Male*			*Female*		
Aero.	40	173	114	3	30	23	1	7	9	0	0	0
Civil	0	0	0	0	0	0	0	7	0	0	2	0
Comp. Sci.	70	41	35	25	5	8	9	22	65	5	17	16
E + E	12	27	8	1	4	2	2	13	21	0	3	3
Eng. Math.	0	0	0	0	0	3	0	0	1	0	0	0
Mech.	3	5	4	0	0	0	0	7	0	0	2	0
TOTAL	125	246	169	29	39	36	12	56	96	5	24	19

Table A4.12 Intake targets for Home Undergraduates for the courses offered by the Faculty in the sessions 2005–07. 'Euro' attached to the course indicates that the third year (of four) is spent in a European Institution

	2005	*2006*	*2007*
Aero. Eng	46	46	49
Aero./Euro	10	10	12
Avionics	12	12	7
Civil Eng.	40	40	44
Civil/Euro	15	15	16
Comp. Sys.	17	17	10
Comp. MSci	25	25	30
Comp. Sci.	35	35	35
Elec. + Com Eng.	16	16	13
Elec. MEng	36	36	30
Elec./Euro.	5	5	3
ElecEng	2	2	2
Eng. Design	25	25	25
Eng. Math.	20	20	21
Mech. Eng.	60	60	65
Mech./Euro	5	5	5
TOTAL	369	369	369

Table A4.13 Faculty courses offered in 2003 together with entry statistics

ENTRY STATISTICS 2003						
Programme	UCAS code	Places	Applicants	Offers	Av. A-level score (max. 360)*	Length (years)
Aeronautical Engineering (MEng)	H410	54+	455	253	347.0	4
Aeronautical Engineering with Study in Continental Europe (MEng)	H401		70	39	337.5	4
Avionic Systems (MEng)	HH64	14+	56	46	320.0	4
Avionic Systems (BEng)	HH46		30	5		3
Civil Engineering (MEng)	H200	54+	346	267	335.7	4
Civil Engineering (BEng)	H205		67	18	340.0	3
Civil Engineering with Study in Continental Europe (MEng)	H201		51	43	334.3	4
Computer Science (MEng)	G403	57+	189	130	340.8	4
Computer Science (BSc)	G400		441	206	341.3	3
Computer Science with Study in Continental Europe (MEng)	G401		33	15	340.0	4
Computer Systems Engineering (MEng)	H622	20+	112+	57+	327.3+	4
Computer Systems Engineering with Study in Continental Europe (MEng)	H621					4
Electrical and Electronic Engineering (MEng)	H606	61+	365+	267+	334.1+	4
Electrical and Electronic Engineering (BEng)	H600					3
Electrical and Electronic Engineering with Management (MEng)	H6N2		New programme			4
Electrical and Electronic Engineering with Study in Continental Europe (MEng)	H605					4
Electronic Engineering (BEng)	H610					3
Electronic and Communications Engineering (MEng)	H623					4
Electronic and Communications Engineering (BEng)	H640					3
Communications and Multimedia Engineering (MEng)	H642					4
Engineering Design (MEng)	H150	24	104	60	335.8	5
Engineering Mathematics (MEng)	G161	20+	70+	62+	327.5	4
Engineering Mathematics with Study Abroad (MEng)	G160		New programme			4
Knowledge Engineering (MEng)	G720		New programme			4
Knowledge Engineering with Study Abroad (MEng)	G721		New programme			4
Mechanical Engineering (MEng)	H300	66+	672+	427+	341.4+	4
Mechanical Engineering (BEng)	H305					3
Mechanical Engineering with Study in Continental Europe (MEng)	H301					4

* Average A-level score – maximum 360 points from three A-levels, excluding General Studies.
+ Numbers for the group of programmes indicated.

In Appendix5 the following names should be added
to the main list on page 346.

1979 Ronald J.Bridle BSc 1953 Director R. and D., Min. of Transport.
1979 William A. Gambling BSc 1947 Prof. of Mic.electronics, U.of S'ton.
1980 Anthony R. Flint. PhD 1949. Partner, Flint and Neil Consultants.
1981 Joseph Black Sen. Lect. Aero. Eng.. Prof. of Eng., Univ. of Bath.
1985 Francis Walley BSc 1940 Director, Ministry of Work.
1985 Michael J. Hamlin BSc 1951 Vice-Chancellor, Univ of Dundee.
1987 Thomas V. Lawson. Reader in Aero. Eng., Univ. of Bristol.
1989 Peter B. Morice BSc 1950. Prof. of Civil Eng., Univ of S'ton.
1990 Ronald W. Hobbs. BSc 1944. Consultant, Ove Arup Partnership.
1998 Colin H. Green BSc 1971. Director of Operations, Rolls-Royce plc.
1999 Frederic W. Williams PhD 1964. Prof. of Eng., Univ. of Cardiff.
2002 Stephen W. Huntington BSc 1968, PhD 1973. Chief Executive HR
       Wallingford

Appendix 5

A5.1 Faculty Professors

Those whose start date is 1910 had joined from either UBC or MVTC. Until the mid-1970s each department had only one professor, who was automatically the Head of Department (HoD). Subsequently, as departments grew in size and subject coverage, it became customary to award Personal Chairs to those who had distinguished themselves in research in a particular area – and from 2006 in teaching also, with one of them assuming the role of HoD by appointment for a specified period of years. In the lists below it has been thought useful to note the specific field of each Professor appointed after 1975.

One effect of the greatly increased number of Professors was that their automatic membership of Senate resulted in that body becoming unmanageable; by 1993 it had reached a total of 240. It was then reduced to a total of 100 by a mixture of appointment and election, with attendance as a requirement – failure to do so being recorded in the minutes of the meeting.

Mathematics

This Department changed its name to Theoretical Mechanics in 1950, and to Engineering Mathematics in 1973.

E. S. Boulton	1910–36	
W. M. Shepherd	1959–72	
R. D. Milne	1971–92	HoD 1971–92
J. F. Baldwin	1990–2003	Information technology
S. J. Hogan	1992–	Non-linear mathematics HoD 1992–2004.
A. R. Champneys	2001–	Non-linear applied mathematics HoD 2004–
B. Krauskopf	2003–	ditto
T. P. Martin	2005–	Artificial intelligence
N. Christianini	2005–	ditto

R. R. Clements	2005–07	Engineering education
M di Bernardo	2007–	Non-linear systems and control

Applied Chemistry
This Department was closed on Wertheimer's death in 1923.

J. Wertheimer	1910–23

Geology
This Department was transferred to the Science Faculty in 1949.

S. H. Reynolds	1910–35
A. E. Truman	1936–37
W. F. Whittard	1937–67

Civil Engineering
R. M. Ferrier	1910–28
A. J. Sutton-Pippard	1928–36
J. F. Baker	1936–38
A. G. Pugsley	1944–68
R. T. Severn	1968–96

D. I. Blockley	1989–2006	Risk and systems analysis HoD 1989–94, 2003–06
R. H. J. Sellin	1990–2004	Fluid mechanics HoD 1995–97
D. Muir-Wood	1996–	Soil mechanics HoD 1998–2002
I. Cluckie	1997–	Hydrology and water management
C. A. Taylor	2002–	Earthquake engineering HoD 2007–
P. Godfrey	2005–	Systems engineering part–time industrial professor
S. Gundry	2008	Enterprise and Entrepreneurship

Mechanical Engineering (and Mining until 1927)

J. Munro	1910–20	
A. Robertson	1920–49	
J. L. M. Morrison	1949–71	
G. F. C. Rogers	1964–82	
C. Andrew	1971–82	Applied mechanics HoD 1971–82
E. G. Ellison	1984–95	Fatigue of materials HoD 1985–96
R. D. Adams	1988–05	Vibrations HoD 1996–98
J. Quarini	1992–	Nuclear and process engineering
D. Stoten	1994–	Control engineering HoD 1998–05
D. Smith	1996	Materials
S. Burgess	2005–	Design and nature HoD 2005–
B. Drinkwater	2007–	Ultrasonics

Electrical Engineering (became Electrical and Electronic Eng. in 1973)

D. Robertson	1910–40	
G. H. Rawcliffe	1944–75	
K. F. Sander	1970–84	Electronics HoD 1988–91
B. M. Bird	1979–	Electrical machines HoD 1979–85
E. L. Dagless	1982–	Microelectronics HoD 1985–88
J. P. McGeehan	1988–	Communications engineering HoD 1991–98
S. R. Bowes	1985–2007	
I. White	1995–2001	Optical communications HoD 2000–01
D. Bull	1997–	Signal processing HoD 2001–
A. Nix	2001–	Wireless communication systems
R. Penty	2001–	
J. Rarity	2003–	Optical communication systems
M. Beach	2003–	Radio systems engineering HoD 2006–

A. Bateman	1994–97	Signal processing
P. Mellor	1990–	Electrical Engineering
A. Munro	2003–	Communication networks
C. Railton	1999–	Computational electromagnetics
N. Canagarajah	2004–	Multimedia signal processing

Motor Car Engineering (became Automobile Engineering in 1914; closed in 1933)

| W. Morgan | 1910–33 | |

Aeronautical Engineering (became Aerospace Engineering in 1988)

A. R. Collar	1946–74	
L. F. Crabtree	1975–85	
M. V. Lowson	1986–99	Modern transport systems HoD 1986–99
A. Simpson	1985–	Aeroelasticity HoD 1985–86
S. Fiddes	1997	Aerodynamics
M. Wisnom	1995–	Aerospace structures
M. Friswell	2002–	
N. A. J. Lieven	2002–	Aerodynamics HoD 2002–07
C. B. Allen	2006	Morphing Structures
D. Ewins	2008–	Director of BLADE

(N.B. D. J. Birdsall and M. Lowenburg were non–professorial HoD in 1999–02 and 2008–present, respectively)

Computer Science (Joined the Faculty in 1989 from Science)

M. H. Rogers[1]	1989–95	HoD 1989–95
J. W. Lloyd	1989–98	
D. May	1995–	HoD 1995–04
B. Thomas	1995–2004	Vision systems
P. Flach	1997–	
A. Chalmers	1989–2007	
N. Smart	1990–	HoD 2006–

---

[1] The founder Director of the University Computer Centre in 1963; Prof. of Computer Science in the Science Faculty 1966–89.

C. Melhuish	2005–	
D. Cliff	2006–	
N. Cristianini	2006–	
R. Jozsa	1999–	
D. Pradhan	2001–	
M. Mirmehdi	1997–	Image analysis

Charter plc., Professors in the Principles of Engineering Design – a Royal Academy of Engineering initiative in 1991

Roland Bertudo	Director of Strategic Planning, Rover Group, 1991–97
Chris. Elliott	Director, Smith Systems Engineering, 1991–97
Michael Shears	Director, Ove Arup Partnership, 1991–97
Ted Talbot	Chief Engineer, BAe Airbus, 1991–97
Jeremy Davies	Chief Engineer, Comp. Integrated Eng., Rover Group 1997–
Horst Peter	Man. Dir., ESAB Cutting Systems, Farben, Germany 1997–

A5.2 Deans of the Faculty and Pro-Vice-Chancellors of the University

The first two in the following list were given the title of 'Permanent Dean'. From 1949 to 1997 the term of office was three years, except in the case of illness. In 1997 the Deanship was considered to be a full-time appointment for a period of 5 years and was accompanied by an additional academic appointment to the relevant department.

Deans

1909–24	Julius Wertheimer
1924–49	Andrew Robertson
1949–51	Gordon Hindle Rawcliffe
1951–54	Alfred Grenville Pugsley
1954–57	Arthur Roderick Collar
1957–60	John Lamb Murray Morrison
1960–63	William Morgan Shepherd
1963–66	Gordon Hindle Rawcliffe
1966–69	Gordon Frederick Crichton Rogers
1969–71	William Morgan Shepherd

1971–74	Roy Thomas Severn
1974–77	Kenneth Frederick Sander
1977–80	Colin Andrew
1980–83	Ronald Douglas Milne
1983–85	Louis Frederick Crabtree
1985–88	Brian Michael Bird
1988–91	Eric Dagless
1991–94	Roy Thomas Severn
1994–97	David Ian Blockley
1997–03	Joseph Peter McGeehan
2003–07	David Muir Wood
2007–12	Nicholas Andrew John Lieven

Pro Vice-Chancellors

1961–64	Sir Alfred Grenville Pugsley (Acting Vice-Chancellor 1963)
1966–69	Arthur Roderick Collar (Appointed Vice-Chancellor 1968–9)
1997–80	Gordon Frederick Crichton Rogers
1981–84	Roy Thomas Severn
1990–93	Brian Michael Bird

A5.3 Honours Awarded to Faculty Members

Faculty of Engineering Professors who became Fellows of the Royal Society

1940	Andrew Robertson
1952	Alfred Grenville Pugsley
1954	Alfred John Sutton Pippard
1954	John Fleetwood Baker
1965	Arthur Roderick Collar
1972	Gordon Hindle Rawcliffe
1991	David May

Other Fellows of the Royal Society who were associated with the Faculty

| 1899 | Henry Selby Hele-Shaw. Professor of Engineering at UCB in 1882 |
| 1914 | Thomas Edward Stanton. Professor of Engineering at UCB in 1899 |

1917   Silvanus P Thompson. Professor of Experimental Physics at UCB in 1878

1930   Paul Adrien Maurice Dirac. Electrical Engineering student, 1918–21. Became Lucasian Professor of Mathematics at Cambridge, and won the Nobel Prize for Physics in 1933. Appointed to the Order of Merit in 1973

1943   Sir Reginald Edward Stradling. Civil Engineering student 1909–12. Chief Adviser on Civil Defence at the Home Office during the Second World War, and later Chief Scientific Adviser to the Ministry of Works

1979   Professor Sir Bernard Crossland; Lecturer in Mech. Eng in 1946; Professor of Mech. Eng at Queen's Univ, Belfast 1959, and Chairman of the Enquiry on the King's Cross Tube Station fire

1983   Prof. William Alexander Gambling. Electrical Engineering student in 1943–7, later Professor of Optical Communication at Southampton University

1983   Lord Chilver of Cranfield, undergraduate and postgraduate student, 1943–50. Vice-Chancellor Cranfield Institute of Technology and Chairman of the Milton Keynes Development Corporation

Professors in the Faculty who were elected to the Fellowship of Engineering which was established in 1976, received its Royal Charter in 1983, and became the Royal Academy of Engineering in 1992

1976   Alfred Grenville Pugsley   Founding Fellow
1976   John Lamb Murray Morrison   Founding Fellow
1976   Arthur Roderick Collar   Founding Fellow
1976   Gordon Hindle Rawcliffe   Founding Fellow
1981   Roy Thomas Severn
1984   Brian Michael Bird
1991   Martin Vincent Lowson
1992   David Ian Blockley
1994   Joseph Peter McGeehan
1997   Ian David Cluckie
1998   David Muir-Wood
1998   Patrick S. Godfrey

Other Fellows of the Royal Academy who were associated with the Faculty

1976   John Bernard Caldwell, Founder Fellow. Postgraduate civil engineering student (1950–3). Professor of Naval Architecture, Univ. of Newcastle-upon-Tyne

1977   Lord (Henry) Chilver of Cranfield; student 1943–7; Prof. of Civil Eng. UCL, and Vice–Chancellor, Cranfield Institute of Technology

1979   Sir Bernard Crossland, Lecturer in Mech. Eng 1946, Prof at Queen's, Belfast 1959

1979   Alec Gambling, Elect. Eng. student 1943–7; Prof. of Optoelectronics, Univ. of Southampton

1985   Philip Stanley Bulson. Postgraduate civil engineering student (1950–3). Director Military Vehicles Research Establishment

1987   Alan G. Davenport; Research Fellow 1960–3; Director Boundary Layer Wind Tunnel Lab., Univ. of Western Ontario, Canada

1990   Robert Park, Lecturer 1960–5; Prof. of Civil Engineering, Univ. of Christchurch, New Zealand

2003   Howard S. Wheater; Research Asst. 1975–78; Prof. of Civil Eng., Imperial College

2007   Ian Firth. Civil engineering undergraduate (1982–5). Senior Partner, Flint and Neill, Consulting Engineers

A5.4 Honorary Degrees Awarded by the University to Eminent Engineers

DSc in Engineering
1912   William Wilson
1938   Sir Alfred Hubert Roy Fedden
1951   Sir Archibald Edward Russell
1956   Willis Jackson
1959   Sir Barnes Wallis
1961   Sir Stanley George Hooker
1975   Dietrich Kuchemann
1978   The Lord Baker of Windrush
1981   Sir Angus Paton
1982   The Viscount Caldecote of Bristol

1983   Sir Henry Chilver
1984   Sir George Rowland Jefferson
1987   Donald Leroy Hammond
1987   Gordon Morris Lewis
1988   Admiral Sir Lindsay Bryson

DEng
1990   Sir Robert Reid
1990   John David Wragg
1991   Sir Alan Marshall Muir-Wood
1992   Sir Bernard Crossland
1992   Hugh Metcalfe
1993   John Gordon Collier
1993   Prof. Nigel William Horne
1994   Prof. Jose Medem Sanjuan
1995   David Martin Leakey
1995   Prof. Philip Charles Ruffles
1996   Ronald William Hobbs
1997   Colin Henry Green
1997   John Michael Taylor
1998   Alan Garnett Davenport
1999   Prof. William Alexander Gambling
1999   Sir Robert Wall
2000   Prof. Ralph Benjamin
2000   Prof. John Michael Hamlin
2001   William Ian Liddell
2001   Sir David Robert McMurtry
2002   Prof. Michael Shears
2004   Prof. Patrick Godfrey
2007   Prof. David Milne
2007   Pasquale Pistorio
2008   Peter Head
2009   Peter R.Taylor

MSc in Engineering
1951   Frederick William Partington
1951   Herbert Marston Webb
1951   Sir Charles Lillicrap
1957   Alfred Norman Irens

1960   George Wilfred Goldsmith

Others
1911   D.Sc.   Julius Wertheimer
1949   D.Sc.   Sir Richard Vynne Southwell
1959   D.L.   Andrew Robertson
1966   D.L.   Alfred John Sutton Pippard
1969   D.L.   Arthur Roderick Collar
1994   M.Eng   John Andrew Suthernden Burn

Honorary Fellows of the University – the highest award it can bestow
1949   Em. Prof. Andrew Robertson
1986   Em. Prof. Sir Alfred Pugsley
1986   Em. Prof Roderick Collar – not actually conferred; he died in the week before the Meeting of Court in February 1986

A5.5 The Brunel Lectures, 1984–2009

The Brunel Lecture series was endowed by the late Sir Alfred Pugsley in October 1983, to be given by an eminent Engineer or Engineering Scientist on a subject of current interest to staff and students of the University. Professor Pugsley was Professor of Civil Engineering at Bristol from 1945–68, Pro-Vice-Chancellor from 1961–4, and Acting Vice-Chancellor during the illness of Sir Philip Morris in 1963.

The following are all the Brunel Lectures that have been given:

27 March 1984 (Inaugural Lecture)
Mr Peter Cox of Rendell, Palmer and Tritton – 'The Thames Barrier'.

March 1986
Mr Geoff Myers, Vice-Chairman of the British Railways Board – 'The Future UK Rail System'.

February 1989
Professor John Allen, College of Aeronautics, Cranfield – 'Aerospace in 2038, Achievements and Challenges'.

March 1991
Sir Alan Muir Wood, Sir William Halcrow and Partners –'Channel Tunnel: an Episodical History'.

March 1993
Dr Bob Feilden, Feilden Associates – 'The Contribution of Power Jets Ltd to Jet Propulsion'.

Nov 1995
Professor David Leakey, formerly Chief Engineer for BT – 'Revolutions and Future Shocks in Telecommunications'.

Nov 1998
Martyn Thomas, Chairman Emeritus of Praxis Critical Systems – 'The Millennium Bug – a Consumer Survival Guide'.

Nov 2001
Professor Jim Norton, Chairman of Deutsche Telecom – 'The Future of European e-Business (Real Life after the Dotcom Crash)'.

Nov 2004
Iain Gray, MD of Airbus – 'The future of commercial aerospace – the A380 and beyond'.

Nov 2006
John Roberts, Jacobs Babtie – 'A life of leisure – engineering at theme parks and fairgrounds'.

A5.6 The Arup Lectures, 2002–08 sponsored by the International Firm of Consulting Engineers Arup Associates

2002
Jim Eyre (Wilkinson Eyre, Architects) and Peter Curran (Gifford and Partners) – 'Bridging Art and Science'.

2003
Robin Nicholson (Edward Cullinan, Architects) –'Getting it Together'.

2005
Cecil Balmont (Arup Associates) – 'Informal Networks'.

2006
Peter Head (Director, Ove Arup Partnership) – 'Dongtan – Eco City'.

2007
Herbert Girardet – 'Solar Cities of Tomorrow'.

2009
Chris Luebkeinen – 'The Future is Oversold and Underimagined'.

A5.7 The University College Colston Society Symposia (sponsored by the Faculty)

The Society was founded in 1899 for the promotion of research in the College. From 1949 it devoted the greater part of its income to a Research Symposium held at irregular intervals. Below is the list of these symposia connected with research in the Faculty, and the name of the principal organiser.

1949   Engineering Structures. Prof. A. G. Pugsley, Civil Engineering.

1959   Hypersonic Flow. Prof. A. R. Collar, Aeronautical Engineering.

1969   Communication and Energy in Changing Urban Environments. Prof. D. Jones, Architecture.

1970   Tidal Power and Estuary Management. Profs. R. T. Severn (Civil Eng.), D. L. Dineley (Geology) and L. Hawker (Botany).

2001   Non–linear Mathematics. Prof. S. J. Hogan, Engineering Mathematics.

2006   Adaptive Structures. Drs. D. L. Wagg, I. Bond, P. Weaver and Prof. M. Friswell, Mechanical Engineering.

A5.8 Faculty Ceremonial Officers of the University

W. J. Larnach	Marshall	1988
G. A. L. Reed	Mace-Bearer	1989
R. H. J. Sellin	Bedell	
T. L. Shaw	Bedell	
R. T. Severn	Bedell	
J. W. Smith	Bedell	
P. Warr	Bedell	
R. R. Clements	Bedell	
N. Everitt	Bedell	
R. Haidani-Scott	Bedell	

Appendix 6

Research Assessment Exercises – Rating Criteria

For the first three RAE, 1986, 1989 and 1992

Rating	Description of Quality
5	International excellence in some areas of activity and national excellence in virtually all others.
4	National excellence in virtually all sub-areas of activity, possibly showing some international excellence, or to international level in some and national level in a majority.
3	National excellence in a majority of sub-areas, or to international level in some.
2	National excellence in up to half the sub-areas of activity.
1	National excellence in none of the sub-areas of activity.

For the 1996 and 2001 RAE

Rating	Description of Quality
5*	International excellence in more than half of the research activity submitted and national levels in the remainder.
5	International excellence in up to half of the research activity submitted and national excellence in virtually all the remainder.
4	National excellence in virtually all activity submitted, possibly showing some evidence of international excellence.
3a	National excellence in over two-thirds of the activity submitted, possibly showing evidence of international excellence.
3b	National excellence in more than half of the research submitted.
2	National excellence in up to half of the activity submitted.
1	National excellence in none, or virtually none of the activity submitted.

For the 2008 RAE

Rating	Description of Quality
4*	World leading in terms of originality, significance and rigour.
3*	Internationally excellent in terms of originality, significance and rigour but which nonetheless falls short of the highest standards of excellence.
2*	Recognised internationally in terms of originality, significance and rigour.
1*	Recognised nationally in terms of originality, significance and rigour.
Unclassified	Below standard of nationally recognised work. Or does not meet the published definition of research for the purposes of this assessment.

References

1   McGrath P.
The Merchant Venturers of Bristol; a History of the Society of
Merchant Venturers of the City of Bristol from its Origin to the
Present Day. Pitman Press, Bath, 1975.

2   Timoshenko S. P.
History of Strength of Materials, McGraw Hill, 1953.

3   Macqueen J. G. and Taylor S. W.
University and Community: Essays to mark the Centenary of the
Founding of University College, Bristol. University of Bristol Press,
1976.

4   The Finniston Report
Report of the Committee of Enquiry into the Engineering Profession
3417 HMSO, October 1967. Cmnd 7794 January 1980.

5   'Education for Industry and Commerce' – the 1950 Report of the
National Advisory Council (NAC), together with the UGC
Comments thereon - 'A Note on Technology in Universities'.

6   Fort A.
PROF – The Life of Frederick Lindemann, Pimlico Press, 2004.

7   The Robbins Report
The Future of Higher Education in the UK – The Robbins Report.
Cmnd, 2154 HMSO, October 1963.

8   The Brain Drain Report
'The Brain Drain' – Report of the Working Group on Migration,
Chairman Dr. F. E. Jones. Cmnd 3417, HMSO, October 1967.

9   The Hale Report
      The Hale Committee Report 'On University Teaching Methods',
      HMSO, 1964.

10   The Dainton Report
      The Dainton Report – Enquiry into the flow of Candidates in Science
      and Technology into Higher Education. Cmnd 3541, HMSO,
      February 1968

11   The Development of Higher Education into the 1990s. A Govt. Green
      Paper, Cmnd 9524, HMSO, May 1995.

12   CVCP Report on ' The European Community and Higher Education',
      1972.

13   The Chilver Report
      The Education and Training of Civil Engineers. A Report by the
      Institution of Civil Engineers, London, 1975.
      N.B. The IMechE published a similar report (Dawson), as did the IEE
      (Merriman) at about the same time.

14   The Jarratt Report of the Steering Committee for Efficiency Studies
      in Universities. A Report to the Committee of Vice-Chancellors and
      Principals (CVCP), March 1985.

15   The Dearing Report
      Report of the National Committee of Enquiry into Higher Education
      – Higher Education in the Learning Society. July, 1997. (Govt.
      response in Green Paper 'The Learning Age', Feb 1998.)

16   'The Future of Higher Education.' A paper by the Secretary of State
      for Education and Skills. Cm. 5735, Jan 2003.

17   Govt. White Paper 'Realising our Potential – A Strategy for Science,
      Engineering and Technology.' Cm. 2250, May 1993.

18   Royal Society Report
      A Higher Degree of Concern, Twenty-ten and Beyond, Policy
      Document 02/08, January 2008.

19   CVCP Report
      'Costing, Pricing and Valuing Research and other Projects' – Report
      and Recommendations to Universities, March 1998.

20   Chapman C. R. and Levy J.
      An Engine for Change – the Chronicle of the Engineering Council,
      2004. ISBN 1-898126-64-x.

21   Crossland B.
      The Anatomy of an Engineer – An Autobiography. N. E. Consulting,
      2006.

Indexes

Because Chapters 11–16 give a chronological ordering of principal Departmental events and those who took part in them, names of people and organizations appearing there have been omitted from this Index. This is true also for pages i–xxxvi.

Index of People

Index of Faculty Oganisations

(R.C. – Research Centre)

Index of University Faculties (F) and Departments (D)

Index of University Organisations

Index of other Universities

Index of Colleges (C) and Schools (S)

Index of Learned Societies and Research Organisations

Index of Professional, Commercial and other Organisations